Cities Responding to Climate Change

Stephen Jones

Cities Responding to Climate Change

Copenhagen, Stockholm and Tokyo

palgrave
macmillan

Stephen Jones
University of Queensland
Coorparoo, QLD, Australia

ISBN 978-3-319-64809-5 ISBN 978-3-319-64810-1 (eBook)
DOI 10.1007/978-3-319-64810-1

Library of Congress Control Number: 2017955055

Cover credit: marungaman/Getty

Printed on acid-free paper

This Palgrave Macmillan imprint is published by Springer Nature
The registered company is Springer International Publishing AG
The registered company address is: Gewerbestrasse 11, 6330 Cham, Switzerland

For Anna

Acknowledgments

A book on climate change policy by city government must attempt to capture the essence of context and its role in shaping responses, and this is particularly so in a book with an international focus. I received help from people along the way, and I owe a great debt of gratitude to specific individuals who influenced my thinking with their perspectives and experience.

A first acknowledgement must go to my home institution, the University of Queensland's Business School. Over the years it has supported my research efforts and, more specifically, enabled me to spend time working on completing the book.

A second acknowledgment must go to my friend and colleague Professor Geert Bouckaert, from the Institute for Public Governance at KU Leuven. Our many discussions on the nature and pervasiveness of performance management shaped my ideas on the topic and much of the approach taken in the book.

Third, I must acknowledge Professor Brian Head, from the School of Political Science and International Studies at the University of Queeensland, for his friendship and guidance at those critical times when I needed a wise and deeply analytical mind to show me the way through.

A fourth acknowledgment must go to Rachael Ballard, Head of Science and Society at Palgrave Macmillan/Springer Nature, for her support. She immediately recognised the potential of the book and smoothed the path for me throughout the process. I must also acknowledge Jack Redden, Rachael's Editorial Assistant, for easing me through the final stages toward publication.

I have a great deal of respect for the senior managers from the city governments in Copenhagen, Stockholm and Tokyo who face the political and administrative challenges of responding to climate change. They were generous with their time and their information in outlining their experiences for me. I thank them heartily for their honesty and commitment.

My deepest gratitude is for my partner and best friend Anna. She inspires me and supports me in ways that continue to astound and amaze; this is why the book is dedicated to her.

Contents

List of Figures

List of Tables

List of Boxes

1

Introduction

We build up the world's great cities.
And out of a fabulous story
We fashion an empire's glory:
One man with a dream, at pleasure….
—Ode, Arthur O'Shaughnessy

Though hundreds of thousands had done their very best to disfigure the small piece of land on which they were crowded together, paving the ground with stones, scraping away every sprouting blade of grass, lopping off trees, driving away birds and beasts, filling the air with smoke of coal and oil,—still spring was spring, even in the town.
It was not this spring morning that they thought sacred and important, not the beauty of God's world, given to benefit all creatures—a beauty which inclines the heart to peace, harmony, and love—but their own devices for getting the upper hand of each other.
—Leo Tolstoy, 1899. Resurrection, 1–2

Introduction

Jaccard (2016) claims 'the thing about effective climate policy is that it is never a political winner'. This simple statement summarises a complex situation. In the policy literature, a successful policy is described as one

S. Jones, *Cities Responding to Climate Change*,
DOI 10.1007/978-3-319-64810-1_1

that achieves the policy-maker's objectives. McConnell (2010) regards policy success as the condition when a policy 'achieves the goals that proponents set out to achieve and attracts no criticism of any significance and/or support is virtually universal' (p. 351). The health policy that reduces the incidence of smoking in young people, the traffic management policy that reduces congestion—these are common examples. One could argue that climate mitigation policies are effective if they reduce greenhouse gas (GHG) emissions in accordance with the targets set by governments. Adaptation policies are successful and effective if they protect areas vulnerable to extreme weather events. On a broader, more complex level, climate policies are effective if they change damaging unsustainable practices into safe sustainable ones. Effective policies show demonstrable results, and policy-makers can provide the details on how the policy has worked and how it achieved the government's intentions. There is accountability and transparency in a democratic political system, where policy-makers demonstrate the effectiveness of their policies in ways that are clear and understandable to citizens. The policy literature also stresses that public policy is not always straightforward. The reality of public policy success is far more nuanced and frustratingly less observable; in fact, policy is a 'complex phenomenon consisting of numerous decisions and made by numerous individuals and organisations' (Howlett et al. 2009, 7). On this basis alone, it may be more difficult to identify reasons why policies are effective. What is clear from the literature is that public policy requires ongoing study and analysis in the hope that we can continue to do it better (Wheelan 2011).

The second part of Jaccard's statement hints at factors that may prevent climate policy from being effective. Policies may be regarded as a failure when they do not achieve the goals the proponents set out to achieve and no longer receive support from them (McConnell 2010). The experiences of governments developing and implementing climate policy have been fraught with challenges that prevent success, so much so that it is regarded as the 'preeminent policy challenge of our time' (Bransetter and Pizer 2012, 3). The policy implementation literature explains that policy-makers often do not foresee many of the barriers that will prevent their policies from being successful and effective (Sabatier and Mazmanian 1979). The challenges highlight the complexity of public policy and the need for an in-depth understanding of the factors involved in determining success or failure. Sometimes, barriers are institutional; for example, another level of government can override the policy, effectively ruling out any chance of success. In some cases, a government has not done the relevant research to show what factors will impact on the implementation process or how their policy decisions will effectively deal with the issues they are targeting. In other instances, those with the task of implementing the policy have their own biases that can effectively sabotage the policy. Implementation of the policy can also be thwarted by resistance in the community and from pressure groups, particularly those who are negatively impacted by the policy. These challenges continue to impact on both the development and the implementation of climate policy by all levels of government. Since the 1990s, many city governments have been attempting to improve the effectiveness of their policy action through the application of performance-based approaches. These approaches focus on measuring their actions in the context of policy objectives and producing public reports that outline their effectiveness. The idea being that measurement and reporting will help to identify the problems and challenges for policy effectiveness and provide the necessary evidence for improvements to future action.

Performance management approaches that require the tracking and measurement of strategic goals, targets, and achievements are seen as the most widespread governmental reform in recent decades (Moynihan and Pandey 2010). Performance management holds the potential to improve the effectiveness of climate policy because it promises to enable governments to quantify promises, and measure and report the results of their actions in ways that allow citizens, managers, and politicians to make meaningful decisions about increasingly complex government activities (OECD 2009). Embedded within performance management is the assumption that measurement and reporting will also drive improvements in the efficiency and effectiveness of government organisations (Dubnick 2005). A key assumption is that performance data can be used to better allocate resources, make decisions about strategy, re-engineer processes, motivate workers, and usher in a new era of accountability (Moynihan 2009). The promise is that measurement—the quantitative representation of the quality or quantity of inputs, outputs, and/or outcomes

(Sterck and Scheers 2006)—can offer valuable data for policy-makers by assisting with comparison, prediction, and evaluation. Importantly for policy-makers, there appears to be a link between performance reporting and trust. If citizens find performance information from government agencies to be credible, they are more likely to believe what is being reported (James and Van Ryzin 2017). This acceptance is critical for climate change policy, as performance information should ideally contribute to the changes necessary to improve policy action by city governments towards more effective environmental, economic, and political sustainability.

The objective of this book is to establish an analytical framework to examine the application of performance management by city governments in the context of their policy response to climate change. This examination will lead to conclusions that can provide guidance for other cities seeking to implement successful and effective policies when responding to climate change. The promise of performance management and its contribution to improving the effectiveness of climate policy is a serious issue that warrants investigation. City governments have been making commitments to respond to climate change, with many arguing that they can demonstrate the success of their efforts. There have, however, been questions raised in the policy literature about the seriousness with which city governments have been willing to measure and report on the results of their policy action. Current assessments suggest that the extent to which city governments engage with performance management has been mixed and uneven (Ammons and Roenigk 2015). The analytical framework will be used to examine cases of city governments committed to performance management systems when measuring the effectiveness of their response. A case study approach provides valuable information for both researchers and policy practitioners, as it is an in-depth consideration of context, different perspectives, and the evaluation of various factors that help explain diversity (Rhodes et al. 2012). The case studies will progress the study of performance measurement and management through these factors and attempt to determine how they influence results (van Thiel and Leeuw 2002; Spekle and Verbeeten 2014). The book is therefore a grounded exploration of city governments and their application of performance to improve the effectiveness of their climate policies. The analytical framework goes beyond current approaches to the examination of performance by city governments responding to climate change. The framework will be used as the guiding instrument for the case studies of city governments and will focus on the dimensions of performance management that include clarity of climate goals, climate targets and related measurement systems, the extent to which performance is embedded within their management and governance processes, and how performance information is used to report progress and encourage management and policy change. This introductory chapter will begin by briefly discussing the contextual issues linking climate policy and performance management that will be covered in greater depth throughout the remainder of the book.

Context

In December 2016, the *Financial Times* reported that over 7000 cities and towns from 119 countries had signed up to join forces to fight climate change. These cities have agreed to work on policy initiatives to reduce GHG emissions and to make cities more resilient to the potential impact of global warming (Crooks 2016). Their agreement is based on assumptions that city governments need to take action because national governments have failed to respond in an effective manner. According to city governments, the leaders of national governments have been failing to understand that their citizens want more effective responses to climate change (ICLEI 2009). City leaders argue that public support is driving the issue, and that without effective action, citizens are concerned about the long-term impact of climate change (Lithicum 2016). Government policy is a complex business because it should reflect community values. Problems in the majority of policy areas will emerge if policy-makers fail to understand these values or fail to clearly articulate why some values are more important at particular times or in particular instances, for example, development is more important than conservation. City governments argue that they understand this and are prepared to make the decisions to reflect the values of their communities when responding to climate change (OECD 2015). According to many city leaders, national governments have lost sight of community values and now represent the values of industry and business. More particularly, many cities point to arguments about high levels of government subsidies as evidence of policy capture by the fossil fuel industry, which prevents effective action to deal with the challenges of climate change (Bast et al. 2015).

Cities contain high concentrations of consumers who need large amounts of energy to maintain their high and unsustainable standards of living. Current estimates by the United Nations (UN) show that cities account for more than 60% of global energy use, 70% of GHG emissions, and 70% of global waste. Current practices are depleting the Earth's finite resources, changing its climate, and damaging its natural ecosystems (Stephan et al. 2016; Gouldson et al. 2015; Dodman 2009; Satterthwaite 2008). It is because cities are so prominent that some argue that they also provide the source of the solutions to take serious action in responding to climate change. Improvements to energy efficiency, improved building standards, better planning to reduce urban sprawl and traffic congestion are part of the New Urban Agenda, which seeks to change the consumption patterns and improve the efficiency of our cities (UN 2016). Following the COP22 Climate Change Conference in Marrakech, Morocco, the Swedish Minister for the Environment, Karolina Skog, echoed the UN view when she claimed 'cities now have the opportunity to demonstrate leadership in transitioning to sustainable societies the world over' (Sweden 2016, 2). These are wonderful sentiments, but the majority of city governments in their individual contexts cannot easily make these transitions without the support of other levels of government.

In many cases, city governments have a limited range of powers—'roads, rates, and rubbish' is the common and rather dated idiom. Cities also face resource restrictions unless their institutional arrangements give them access to their own sources of revenue, for example, income taxes. Otherwise, many city governments are reliant on grants and tied funding arrangements from other levels of government. Policies affecting settlement patterns, energy production, and emissions are often joint responsibilities. The policy levers that impact on these areas can be the responsibility of national governments. Some writers claim that without joined-up collaborative approaches, involving all levels of government, climate policy at the city level may never be effective (Kern and Abler 2008; Betsill and Bulkeley 2006). The cities examined in this book will provide details on these institutional factors, and the opportunities and constraints they place on city governments to develop and implement climate policies.

City governments can claim a serious intent to respond to climate issues, but in effect are often unable to take substantial action. Many city governments have been establishing climate policies that are well beyond their legislated capacity. For example, the capacity of many city governments to implement policies that explicitly target the energy consumption of individual dwellings, such as energy efficiency standards or passive solar design, is limited (Betsill and Bulkeley 2004). There are various reasons for this limitation, ranging from lack of resources, making decisions in areas not in their control to overly ambitious politicians. Each of the case studies in this book examines the range of climate policies established by each level of government relevant to each city. In each instance, the case attempts to link the local climate action to the wider institutional context to explain the consistency across levels of government. This linkage will help to reveal the ambitions of the city and the potential conflicts posed by national government action or inaction.

Climate change is a policy area that brings together substantial issues affecting cities and the governments that run them. Discussions about climate change focus on the environment and sustainability, and on the need to reduce human impact. These discussions invariably involve consideration of economic issues and how actions taken to respond to climate change must be undertaken with minimal or no cost (Giddens 2011; Parr 2014). This will be an important component of the case study analysis. Governments need to decide if they will adopt a legislative approach and introduce mandatory regulations to bring about the changes needed to reduce human impact or just present convincing arguments and make recommendations to encourage change. Questions then emerge on the most appropriate policy process to adopt. Should the government take a top-down or a bottom-up approach; who should be consulted; which interests take precedence? Since the 1990s, neoliberal models have dominated administrative and management approaches taken by many city governments. The measurement and reporting of performance has been a widely applied aspect of the neoliberal approach as a solution for city governments facing public demand to demonstrate their efficiency and effectiveness in delivering services to their communities. The case studies examined here will reflect on these dominant governance models and consider the most effective ways to deliver services that have been challenging local decision-makers.

Survey data reveals a high degree of consistency to the voluntary approaches taken by local governments in the United States to climate change policy (Kraus 2011). The main approaches to mitigating GHG emissions taken by US cities tend to focus primarily on recycling of waste and the provision of information for residents on energy efficiency. Research has been quite clear for some time that voluntary approaches targeting production processes appear not to generate significant pollution abatement (Koehler 2007). Research by Kern and Abler also found that in both mitigation and adaptation initiatives, city governments 'do not fully exploit their authoritative powers and are reluctant to apply authoritative modes of governing through regulatory measures and strategic planning' (2008, 171). Of course, as these authors found, the capacity to implement climate change policies is closely linked with cities' regulatory modes of urban governance (ibid.). Governments are reluctant to propose mandatory measures because of the potential pushback from business and industry. The case study cities in this book will be asked to provide some insights into this issue. The literature suggests that many city governments believe that if they apply policy tools such as taxes, emissions trading or regulatory standards will undermine 'competitiveness, especially if similar regulations are not required of their international competitors' (Kern and Abler 2008, 60). Australian and Netherlands research on voluntary approaches in building sectors increasingly demonstrates non-regulatory approaches' limited effectiveness in engaging industries on climate and energy issues (Van der Heijden 2015). The most recent international comparative research finds that there may be a role for mixed approaches and emphasises the complementary potential of voluntary and regulatory approaches to advancing energy efficiency and climate resilience (Trencher et al. 2016). The responses provided by the case studies will extend our current understanding of the choices behind the approaches taken.

When city governments consider establishing a performance system to measure the effectiveness of their climate policy, there are at least two challenges they need to address: first, to identify ways to localise measurements on climate change, and second, to establish standardised measures that are comparative, allowing for benchmarking through common methodologies (McCarney 2009, 29). The case studies in this book will examine city governments' responses to these challenges by asking whether they can actually demonstrate, through objective measurement practices, effectiveness in achieving climate objectives. Demonstrable success is important for two fundamental reasons. First, cities within their individual context can fulfil a role as sources of experimentation and innovation (Krause 2011). City governments can be more agile in decision-making than national governments and are closer to their citizens. Being quick in responding to climate issues with demonstrable results can provide a level of public support and evidence for other levels of government to pursue policies. By acting as pilot sites, cities can help provide the evidence of what works in terms of responding to climate change. Establishing performance measurement indicators can help city governments to build more effective planning and efficient management for climate action.

The second reason relates to the international context where city governments have established networks of collaboration such as Local Governments for Sustainability (founded in 1990 as the International Council for Local Environmental Initiatives [ICLEI]), C40, and the Global Covenant of Mayors for Climate and Energy to respond to climate change (Osofsky and Levitt 2007). Evidence of success is a major component of the agenda of these networks to be taken as equal partners in international negotiations on climate goals (Bulkeley and Broto 2013). Performance indicators on climate change at the city level can inform city officials and support their existing, and indeed potentially far-reaching, powers of planning, aimed at climate change adaptation and mitigation (McCarney 2009, 28). Performance and measurement of results are therefore critical to both national and international contexts. Rigorous data-driven policy analysis by cities could strengthen their position in intergovernmental relations and decisions on climate change action. In addition to the contextual challenges outlined above, the application of performance management brings with it a complexity that also challenges city government political representatives and their administrators.

Performance management has the potential to assist city governments facing the policy and governance challenges associated with climate policy. City governments will not however become successful in performance management overnight. Being good at performance management takes time. If it requires cultural change, it will take a long time—some suggest up to 10 years (McDonald et al. 2003). New attitudes towards how work is assessed have to be developed, new routines introduced and learned, new work practices and action programmes designed, and new relationships with internal and external stakeholders formed. Julnes and Holzer (2001) examined the unique challenges of utilising performance measurement in public organisations. They identified both rational/technocratic factors and political/cultural factors; their findings are summarised in Table 1.1. The findings were reasonably clear in concluding that the adoption of performance as a driver of change is influenced by rational/technocratic factors such as information, resources, goal orientation, and so on. Implementation is more influenced by political and cultural factors that are difficult to change such as attitudes, interest groups, and unions. Common across both factors is limited or poor-quality information and a lack of resources committed to the gathering and processing of performance measurement data.

The successful application of performance management requires many obstacles to be overcome. Understanding these challenges through the

Table 1.1 Factors affecting utilisation of performance measurement in public organisations

	Adoption	Implementation
Rational/Technocratic	External requirements Internal requirements Resources Goal orientation Information	Resources Information
Political/Cultural	Internal interest groups	External interest groups Degree of unionisation Risk-taking Attitudes

Source: Julnes and Holzer (2001, 702)

experiences of others may help city leaders to establish a process leading to ongoing improvement. Performance management holds significant promise for improvement in the implementation of climate policy that suits both the local and international agendas for city government action. One of the key objectives of this book is to look for ways in which city governments can improve their application of performance management in their response to climate change. The intention is to establish a staged approach to help city leaders identify what point a city government has reached on progressing to more effective and successful application of performance management principles that help to improve policy effectiveness. Staged approaches have been used in disciplines such as psychology for the study of group dynamics and the study of business organisations. In the public sector, staged approaches have been applied to service innovation and improvements to implementation (Walker 2003; Schofield 2004). By establishing a staged approach, the expectation is that city leaders can identify what challenges they face in advancing performance management within the management and administration of municipal climate action. This will help to clarify potential confusion about levels of progress, identify what additional actions could to be taken, and set benchmarks and goals in taking appropriate actions to advance to the next stage.

The next section will provide introductory details on other aspects of performance management and the challenges it presents.

Challenges in Measuring Performance

The foundation of an effective measurement system is access to good-quality data. For many city governments, this is where the challenge of climate policy begins, and often ends. The baseline for measuring mitigation policies is based on the UN protocol, and many cities have established inventories based on the simple approach developed by the ICLEI network. However, there are limitations to such systems, and researchers examining the protocols behind these inventories have been raising questions about their accuracy, comparability, and coverage. Closer inspection shows that there are several technical issues with GHG inventories for cities, including the lack of a full life-cycle perspective covering all aspects of production. Problems also arise when defining spatial and temporal context (e.g., exactly where are the boundaries of any given city?). Finally, there are complicating issues around assigning emissions by political jurisdiction. This is particularly problematic within multilevel systems, where different governments have responsibility for different activities that cause emissions (Kennedy et al. 2010). Research in this area breaks emissions down into three 'scopes' that represent possible sources within the city context. Scope 1 covers emissions within municipal boundaries and includes fossil fuel combustion, waste (landfill) emissions, industrial processes and product use, and emissions from agriculture, forestry, and other land use. Scope 2 focuses on emissions from power plants within the municipal boundary. Scope 3 emissions are the most challenging for a single level of government to deal with, as they include not only production-based emissions but also those from food and materials consumed within the city (ibid.).

The main message from the literature on inventories is that there are multiple sources for overall emissions in any given city. Clarity on the sources and their volumes will be a challenging task. It seems reasonably clear that when city governments begin to establish a baseline inventory as the foundation of a performance-based approach, they have at least three choices. First, they can rely on their national government to measure emissions. This approach may or may not break emission sources down to the municipal level. It will depend on what the national government chooses to measure and will be unlikely to determine accurately the results of the city's individual policies intended to reduce emissions. Second, a city government can develop its own inventory based on local knowledge using its own resources. This approach may give it a reasonably accurate picture of the emissions produced within its boundaries. This will not, however, give it the opportunity to accurately compare how effective its efforts are because it does not have an accurate basis for comparison, and opportunities to learn from other cities will be limited. The third choice is that city governments have access to methodologies primarily developed by non-government agencies, which are networks of other city governments trying to do the same thing. This will allow them to work with the network on identifying more effective ways to reduce emissions. One interesting aspect of the understanding within these city-based networks is that policy learning should take place and there is a focus on identifying cities with better per capita performance than others.

In each of the three baseline inventory choices available to cities, GHG inventories will be based primarily on the production of GHG emissions and not consumption-based emissions. On this basis, city governments are focused on one part of their total GHG emissions, and the more developed the city, the more likely that consumption-based emissions are larger than those emanating from production (Mi et al. 2016). As researchers identify improved methodologies that also identify consumption-based emissions, and means to reduce them, there may be possibilities, with the appropriate political will, that those city government climate policies will focus on a more complete set of emissions. This development will help move cities to a point of greater accuracy in their policies to mitigate climate impacts.

Another aspect to establishing a performance-based system is determining which indicators are most appropriate to show progress. City governments face major challenges in this area because the quality of the indicators measures the quality of the performance. Most importantly for governments, the indicators are a way of ensuring accountability, as they are the 'currency' of evaluation (Carter 1995). While indicators are only one part of a performance scheme, they are critical because they determine what is measured and what is reported. Indicators consequently tell us what good performance is and whether city governments are actually effective in responding to climate change. The focus on the contribution of indicators to fulfil measurement, reporting, and verification was established in the commitments of parties under the United Nations Framework Convention on Climate Change (UNFCCC) in the Bali Action Plan in 2007. The decision set in motion the international agenda for the development of

> measurable, reportable and verifiable nationally appropriate mitigation commitments or actions, including quantified emission limitation and reduction objectives, while ensuring the comparability of efforts among them, taking into account differences in their national circumstances. (UNFCCC 2007)

Indicators play an important and positive role in policy-making, as they can provide early and timely reporting of progress and setbacks, allowing for policy learning and discussion of best practices across projects or programmes. Developing good indicators is not easy, particularly for governments, as there are often unclear indicators of success. Climate change is commonly referred to as a wicked problem—problems that, by their nature, have innumerable causes, are difficult to describe, can involve incomplete or contradictory knowledge, are subject to different opinions, and are interconnected with other problems. There is no single solution to a wicked problem, so such problems need multidisciplinary approaches, which often have their own methodologies and ways of determining outcomes. These problems require wide-scale approaches that consider a range of variables. The World Bank looks at it this way:

> Climate change is an issue that presents great scientific and economic complexity, some very deep uncertainties, profound ethical issues, and even lack of agreement on what the problem is. Economists will generally think about the trade-offs involved. Ecologists will talk about the idea that we're driving towards the edge of a cliff. The question is, how do you reconcile these two—if you can? (World Bank 2014)

Yet the indicators that are chosen to measure progress will often be influenced by how we interpret the problem. An economist will propose indicators that consider financial burden; an ecologist will be concerned about environmental impacts; a politician may be concerned about electoral consequences. In an ideal situation, when establishing a performance scheme, city leaders need to consider which indicators reflect the interests and concerns of their community. Metrics should be appropriate, relevant, selective, simplified and outcome oriented, and able to capture cross-cutting outcomes (Cust 2009). Performance budgeting, cost–benefit analysis, and management by objectives have been common managerial responses to such dilemmas (Hwang 2016; Adams et al. 2014). The idea being that decision-makers can be provided with the evidence that shows the most pressing problem and the most appropriate response to deal with it. The assumption is that following these managerial approaches will introduce some rationality into politically and emotional charged issues (Head and Alford 2015). They also provide guidance on establishing the most appropriate indicators to measure performance. Common indicators should ideally be aligned with common inventory methods. Valuable work by Cust (2009) emphasises the importance of indicators in the policy implementation process:

> Informative and policy-relevant indicators, once successfully designed and implemented, can facilitate better policy design, ongoing assessment and updating—generic indicators are of significantly less value. The learning and performance-management benefits of quantitative indicator sets are further enhanced where data are available to stakeholders, or where indicator data are actively shared across domains. (Cust 2009, 461)

Another significant challenge for city governments when adopting a performance-based approach is the 'use' to which they apply the performance data they gather. The performance management literature shows a high level of interest in this aspect of the policy implementation process. In terms of the prime use of performance information, Behn (2002) argues that public managers' real purpose—indeed the only real purpose—is to improve performance. Benchmarking and other forms of comparison can be an effective analytical use of performance data in each of these key purposes. Comparisons with past performance, with other organisations in the sector, or with international standards can all contribute to the use of information for the purpose of improving performance. There are ongoing debates in the policy literature on the effective use of performance information. The current view is that public organisations devote significant resources to creating performance information systems; however; they largely neglect the question of how to encourage the use of the information for the purposes of management and policy change. Some of the leading research in this area suggests that whether managers and other employees in public organisations use performance information is influenced by the social context and the formal system in which they work (Moynihan and Pandey 2010). Leadership/Political support for performance management matters, goal-oriented cultures matter, and citizens' support for and involvement in performance management processes matters (Saliterer and Korac 2013; Bjornholt and Larsen 2014).

In addition to these established understandings, recent research also shows that access to and integration of information into performance management systems predict greater use. It is also important for organisations to foster having an open, innovative, and risk-taking culture where staff are encouraged to learn from new data that supports the need for change (Moynihan and Pandey 2010). The innovative approach can provide one of the most difficult challenges for public organisations (Sanger 2013). Government administrations that are based on the traditional bureaucratic culture that stresses hierarchy, process, and procedure over flexibility and risk-taking have been shown to resist the application of performance as a basis for management (Parker and Bradley 2000; Sharma 2005). City governments need to be aware that the use of information is both a supply and a demand situation. The collection of performance information can establish the necessary supply, but organisational factors must also create the demand for its use.

The challenges outlined in this brief introduction provide important insights for city governments relating to their use of performance information to improve effectiveness in the implementation of climate policies. How these governments embed performance into management processes stands as an indicator of a commitment to performance as a tool to support accountability. 'Use' of performance measurement data allows for internal and external comparison, which in turn supports policy learning and improvement. Bouckaert and Halligan (2008) argue that 'incorporation' and 'use' of performance information are critical to the effectiveness of a performance management system. It is important to recognise the importance of 'use' as the critical element in the performance agenda, claiming, 'if we want to study the successes and failures of performance movements, we have to study the use of performance information' (Van Dooren et al. 2008, 22). Reporting of performance is a critical use of information and a means of providing an account to internal and external stakeholders of current and past performance. Reporting requirements can be imposed on organisations, by political leaders, or by another level of government; in most cases, optimal indicators focus reporting on the outcomes of policy initiatives. In the context of the theme of this book, it is worth noting that researchers have also found that there is potential for reporting or benchmarking programmes to later transition to models mandating performance improvements, such as 'cap and trade' (Trencher et al. 2016, 353). Such observations stress the necessity to examine the relationship between policy implementation and the application of performance management. External accountability demands high levels of validity and reliability in the reported information on the effectiveness of policy action. These demand-side requirements support the learning objectives that should feed back into the ongoing development and supply-side decisions on the development of the measures. The case studies will help to cast some light on these factors and how city governments are responding to these challenges.

Research Strategy

The research strategy requires consideration of the approaches to performance management taken by a wide range of city governments. A telephonic and email survey of cities that are members of the C40 pilot group for establishing the community-based measurement system revealed that few had made anything beyond verbal commitments to the C40 protocol. This early observation is supported by research on the ICLEI network which shows that only 8% of their members undertake any evaluation of the effectiveness of the local climate change strategy (Reams et al. 2012). It seems that remarkably few city governments will establish a climate action plan, mitigation or adaptation, with an implementation process that includes a review and evaluation framework (Wheeler 2008; Preston et al. 2011). With this in mind, there is no attempt in this book to deal with all of the cities discussed at the same level of detail or depth. Throughout the early sections of book, there are examples of approaches to a cross section of city governments in the context of the current literature on climate change and performance management. Various aspects of the cities' performance management system are covered, with some consideration of the strengths and weaknesses of their approaches. Some cities have been active participants in providing feedback to the C40 and the World Resources Institute through their pilot cities programme on various elements of the community-based scheme. It was clear from preliminary discussions with representatives of the C40 that there was a low level of awareness in the city governments that had actually adopted and were applying a measurement methodology. Like other aspects of city network membership, the measurement scheme is voluntary and thus there is no reason to report through formalised mechanisms. Cities that presented clear evidence of effective measurement were disappointingly few. This created a dilemma, as very few city governments responded to approaches by the author to discuss details of the commitment to and participation in performance management.

The cities willing to discuss their approaches in depth enough to build a reasonable case study analysis were Copenhagen, Stockholm, and Tokyo. The characteristics present in these cities are important in understanding the linkages between effective climate policy and performance management. Each city has a strategic planning approach that links a set of climate policies to other activities and responsibilities of the city government. Each has been undertaking climate policies for at least a decade, which provides a substantial period for the implementation of policy. Each has a set of related performance criteria linked to its climate strategy. Each provides public reports outlining its progress in achieving its climate goals. Each is a member of at least one international network of city governments responding to climate change. Each city has been subject to national government decisions that impact on climate policy. Each of the cities had staff members willing to be interviewed to discuss implementation issues relating to climate policies. Copenhagen was chosen because of its comprehensive approach to strategic management that incorporates an extensive and detailed reporting regime. Both factors are important to policy implementation and performance management. Stockholm was chosen for its recognised attempts at mainstreaming policy development and implementation. Again, these are elements identified in the climate policy literature as important to policy implementation. Finally, Tokyo is one of the very few city governments to adopt a mandatory approach to reducing emissions through a city-based cap-and-trade scheme. The scheme has applied performance measurement in policy development and performance as an integral component of its implementation programme. These elements make these city governments important contributors to the examination of performance as a factor in the implementation of climate policy at the city government level.

These cities also had some secondary research on their climate policies, but there had not been a comprehensive assessment of the performance aspect of their approach. The decision to focus on these three cities was made on the basis that findings could underpin future research. By establishing the research methodology and the research framework, these cities would prove to be starting points for building a future research programme and refining some of the substantial work already undertaken in this regard by the author in other cities, for example, Vancouver, Melbourne, and New York City (Jones 2012, 2013, 2015). Based on the experience of researching these cities, the material gathered for this book provides a substantial cross section of lessons for other cities considering the application of performance management to evaluate the effectiveness of their climate policies.

Outline of the Book

This brief introduction to the themes and perspectives on the participation of city governments has raised issues that will require attention and elaboration throughout the remainder of the book. A clear role for city governments in both mitigating and adapting to the impact of climate change is subject to considerable debate. Fundamental questions emerge from the debate and continue to perplex policy-makers and practitioners at the local level. Critical questions include: what are the policy actions these governments need to take to reduce the impact of their decisions on the environment, and what are the most effective policy initiatives they need to take to reduce the impact of global warming and adapt their cities to the consequences? Implementation issues have become significant challenges for city governments pursuing climate change policy. The programmes and procedures in city government management systems will have a considerable impact on the likely achievement of policy outcomes. The measurement and reporting of success will have impacts at the global and local scales. Throughout this book, there are examples of individual governments and networks of cities that illustrate and analyse the issues and arguments outlined above. The studies presented will contribute to further understandings of policy implementation issues and help identify lessons for other governments in building effective processes and procedures.

Chapter 2 examines the nature of city governments' involvement in climate policy. Further consideration of the measurement issues is covered, with some analysis of specific city government's efforts to apply measurement to its climate strategies. The chapter also covers the work of network alliances and the benefits of membership to city governments. Chapter 3 provides an overview of the current situation with regard to climate policy plans established by city governments. Details include issues relating to the development and implementation of plans and initiatives designed to mitigate and adapt to the impacts of global warming. The chapter focuses on those city governments that apply some form of performance measurement system to report on progress. Chapter 4 establishes an analytical framework for performance management as a model based on the widely accepted doctrine of performance management. Dimensions covered include goal clarity, the development of appropriate measurement systems, and the embedding of performance into management practices through a process of incorporation. The use of information in terms of policy and management change, and in reporting the performance against original climate objectives, will also be discussed. There is also some consideration given to the use of information to establish evidence to support behaviour change. The analytical framework contributes to the development of a staged heuristic to assist city governments seeking guidance in areas where improvements need to be made to maximise the benefits of performance management in achieving success in climate policy. The framework forms the basis for the analysis of the case studies in Chaps. 5–7, which provide a detailed individual examination of Copenhagen, Stockholm, and Tokyo city governments, respectively. Each chapter focuses on the experiences of one of these governments and outlines the factors that facilitate and constrain the application of performance management as a means to promote policy and management change. Each chapter provides details from semi-structured interviews with key personnel from each city government and their experiences with the development and implementation of climate policy and the application of performance management principles. Chapter 8 concludes by highlighting the lessons for other city governments contemplating taking serious action through climate policies designed to help mitigate the impacts of global warming. These lessons and their application have profound implications for the effectiveness of future climate policy.

References

Adams, C., Muir, S., & Hoque, Z. (2014). Measurement of Sustainability Performance in the Public Sector. *Sustainability Accounting, Management and Policy Journal, 5*(1), 46–67.

Ammons, D. N., & Roenigk, D. J. (2015). Benchmarking and Inter-organizational Learning in Local Government. *Journal of Public Administration Research and Theory, 25*(1), 309–335.

Bast, E., Doukas, A., Pickard, S., Van der Burg, L., & Whitley, S. (2015). *Empty Promises: G20 Subsidies to Oil, Gas and Coal Production*. Washington: Oil Change International.

Behn, R. (2002). The Psychological Barriers to Performance Management. *Public Performance & Management Review, 26*(1), 5–25. DOI 10.1080/15309576.2002.11643684.

Betsill, M. M., & Bulkeley, H. (2004). Transnational Networks and Global Environmental Governance: The Cities for Climate Protection Program. *International Studies Quarterly, 48*(2), 471–493.

Betsill, M., & Bulkeley, H. (2006). Cities and the Multilevel Governance of Global Climate Change. *Global Governance, 12*, 141–159.

Bjørnholt, B., & Larsen, F. (2014). The Politics of Performance Measurement: 'Evaluation Use as Mediator for Politics'. *Evaluation, 20*(4), 400–411.

Bouckaert, G., & Halligan, J. (2008). *Managing Performance: International Comparisons*. Oxford: Routledge.

Bransetter, L. G., & Pizer, W. A. (2012). *The Climate Change Challenge in a Global Economy*, NBER Working Paper 18214. DOI 10.3386/w18214

Bulkeley, H., & Castán Broto, V. (2013). Urban Governance and Climate Change Experiments. In*Institutional and Social Innovation for Sustainable Urban Development* (pp. 72–87). London: Routledge.

Carter, N. (1995). Performance Indicators: Backseat Driving or "Hands Off" Control. In D. McKevitt & A. Lawton (Eds.), *Public Sector Management: Theory, Critique and Practice*. London: Sage.

Crooks, E. 2016. World's Mayors Vow to Tackle Climate Change at City Level. *Financial Times*, December 2.

Cust, J. (2009). Using Intermediate Indicators: Lessons for Climate Policy. *Climate Policy, 9*(5), 450–463. DOI 10.3763/cpol.2009.0632.

Dodman, D. (2009). Blaming Cities for Climate Change? An Analysis of Urban Greenhouse Gas Emissions Inventories. *Environment and Urbanization, 21*(1), 185–201.

Dubnick, M. (2005). Accountability and the Promise of Performance: In Search of the Mechanisms. *Public Performance and Management Review, 28*(3), 376–417.

Giddens, A. (2011). *The Politics of Climate Change.* Cambridge: Polity Press.

Gouldson, A., Colenbrander, S., Sudmant, A., McAnulla, F., Kerr, N., Sakai, P., Hall, S., Papargyropoulou, E., & Kuylenstierna, J. (2015). Exploring the Economic Case for Climate Action in Cities. *Global Environmental Change, 35*, 93–105.

Head, B. W., & Alford, J. (2015). Wicked Problems: Implications for Public Policy and Management. *Administration & Society, 47*(6), 711–739.

Howlett, M., Ramesh, M., & Perl, A. (2009). *Studying Public Policy: Policy Cycles and Policy Subsystems.* Ontario: Oxford University Press.

Hwang, K. (2016). Cost-benefit Analysis: Its Usage and Critiques. *Journal of Public Affairs, 16*(1), 75–80.

ICLEI (International Council for Local Environmental Initiatives). (2009). ICLEI Statement of Mayors and Local Governments on Failure of International Community to Act on Climate Change. http://www.icleiusa.org/newsevents/press-room/press-releases/iclei-statement-of-mayors-and-local-governments-on-failure-of-international-communityto-act-on-climate-change

Jaccard, M. (2016). B.C.'s Climate Plan Reaches Olympian Heights of Political Cynicism, *The Globe and Mail*, August 21. https://www.theglobeandmail.com/opinion/bcs-climate-plan-reaches-olympian-heights-of-political-cynicism/article31464244/

James, O., & Van Ryzin, G. G. (2017). Incredibly Good Performance: An Experimental Study of Source and Level Effects on the Credibility of Government. *The American Review of Public Administration, 47*(1), 23–35.

Jones, S. (2012). A Tale of Two Cities: Climate Change Policies in Vancouver and Melbourne: Barometers of Cooperative Federalism? *International Journal of Urban and Regional Research, 36*(6), 1242–1267.

Jones, S. (2013). Climate Change Policies of City Governments in Federal Systems: An Analysis of Vancouver, Melbourne and New York City. *Regional Studies, 47*(6), 974–992. DOI 10.1080/00343404.2011.585150.

Jones, S. (2015). 'If You Can Make it Here, You Can Make it Anywhere': Performance Management and PlaNYC Climate Change Initiatives, *Regional Studies.* DOI 10.1080/00343404.2015.1052389.

Julnes, P. D. L., & Holzer, M. (2001). Promoting the Utilization of Performance Measures in Public Organizations: An Empirical Study of Factors Affecting Adoption and Implementation. *Public Administration Review, 61*(6), 693–708.

Kennedy, C., Steinberger, J., Gasson, B., Hansen, Y., Hillman, T., Havrnek, M., Pataki, D., Phdungsilp, A., Ramaswami, A., & Mendez, G. V. (2010). Methodology for Inventorying Greenhouse Gas Emissions from Global Cities. *Energy Policy, 38*(9), 4828–4837.

Kern, K., & Abler, G. (2008). Governing Climate Change in Cities: Modes of Urban Climate Governance in Multi-Level Systems. *Competitive Cities and Climate Change: OECD Conference Proceedings*, OECD, Milan, Italy, p. 171.

Koehler, D. A. (2007). The Effectiveness of Voluntary Environmental Programs—A Policy at a Crossroads? *Policy Studies Journal, 35*(4), 689–722.

Krause, R. M. (2011). Symbolic or Substantive Policy? Measuring the Extent of Local Commitment to Climate Protection. *Environment and Planning C: Government and Policy, 29*(1), 46–62.

Lithicum, K. (2016). 82 Major Cities Are Meeting This Week to Talk About How to Limit the Worst Effects of Climate Change. *Los Angeles Times*, December 1.

McCarney, P. (2009). City Indicators on Climate Change: Implications for Policy Leverage and Governance. *World Bank's 5th Urban Research Symposium Cities and Climate Change: Responding to an Urgent Agenda*, Marseille, France.

McConnell, A. (2010). Policy Success, Policy Failure and Grey Areas In-between. *Journal of Public Policy, 30*(3), 345–362.

McDonald, B., Rogers, P., & Kefford, B. (2003). Teaching People to Fish? Building the Evaluation Capability of Public Sector Organizations. *Evaluation, 9*(1), 9–29.

Mi, Z., Zhang, Y., Guan, D., Shan, Y., Liu, Z., Cong, R., Yuan, X. C., & Wei, Y. M. (2016). Consumption-based Emission Accounting for Chinese Cities. *Applied Energy, 184*, 1073–1081.

Moynihan, D. (2009). *The Dynamics of Performance Management: Constructing Information and Reform*. Washington: Georgetown University Press.

Moynihan, D. P., & Pandey, S. K. (2010). The Big Question for Performance Management: Why Do Managers Use Performance Information? *Journal of Public Administration Research and Theory, 20*(4), 849–866.

OECD. (2015). Fighting Climate Change: What City Mayors Are Doing. OECD Observer Roundtable. http://www.oecd.org/gov/cop21-and-public-governance.htm

OECD (Organisation for Economic Cooperation and Development). (2009). *Measuring Government Activity*. Paris: OECD.

Osofsky, H. M., & Koven Levit, J. (2007). The Scale of Networks: Local Climate Change Coalitions. *Chicago Journal of International Law, 8*(2), 409.

Parker, R., & Bradley, L. (2000). Organisational Culture in the Public Sector: Evidence from Six Organisations. *International Journal of Public Sector Management, 13*(2), 125–141.

Parr, A. (2014). *The Wrath of Capital: Neoliberalism and Climate Change Politics*. New York: Columbia University Press.

Preston, B. L., Westaway, R. M., & Yuen, E. J. (2011). Climate Adaptation Planning in Practice: An Evaluation of Adaptation Plans from Three Developed Nations. *Mitigation and Adaptation Strategies for Global Change, 16*(4), 407–438.

Reams, M. A., Clinton, K. W., & Lam, N. S. (2012). Achievement of Climate Planning Objectives Among US Member Cities of the International Council for Local Environmental Initiatives (ICLEI). *Low Carbon Economy, 3*(4), 137.

Rhodes, M. L., Biondi, L., Gomes, R., Melo, A. I., Ohemeng, F., Perez-Lopez, G., Rossi, A., & Sutiyono, W. (2012). Current State of Public Sector Performance Management in Seven Selected Countries. *International Journal of Productivity and Performance Management, 61*(3), 235–271.

Sabatier, P., & Mazmanian, D. (1979). The Conditions of Effective Implementation: A Guide to Accomplishing Policy Objectives. *Policy Analysis, 5*(4), 481–504.

Saliterer, I., & Korac, S. (2013). Performance Information Use by Politicians and Public Managers for Internal Control and External Accountability Purposes. *Critical Perspectives on Accounting, 24*(7), 502–517.

Sanger, M. B. (2013). Does Measuring Performance Lead to Better Performance? *Journal of Policy Analysis and Management, 32*(1), 185–203.

Satterthwaite, D. (2008). Cities' Contribution to Global Warming: Notes on the Allocation of Greenhouse Gas Emissions. *Environment and Urbanization, 20*(2), 539–549.

Schofield, J. (2004). A Model of Learned Implementation. *Public Administration, 82*(2), 283–308.

Sharma, B. (2005). Local Government Organisation on Its Journey to Becoming a Learning Organisation. *Business Process Management Journal, 11*(4), 388–402.

Spekle, R., & Verbeeten, F. (2014). The Use of Performance Measurement Systems in the Public Sector: Effects on Performance. *Management Accounting Research, 25*, 131–146.

Stephan, A., Trundle, A., Kendal, D., Henderson, H., Kamalipour, H., & Lowe, M. (2016). Our Cities Need to Go on a Resources Diet. *The Conversation*, December 1.

Sterck, M., & Scheers, B. (2006). Trends in Performance Budgeting in Seven OECD Countries. *Public Performance & Management Review, 30*(1), 47–72.

Sweden. (2016). Cities Crucial for Tackling Climate Challenge. Government Offices of Sweden. http://www.government.se/articles/2016/11/cities-crucial-for-tackling-the-climate-challenge/

Trencher, G., Broto, V. C., Takagi, T., Sprigings, Z., Nishida, Y., & Yarime, M. (2016). Innovative Policy Practices to Advance Building Energy Efficiency and Retrofitting: Approaches, Impacts and Challenges in Ten C40 Cities. *Environmental Science & Policy, 66*, 353–365.

UN. (2016). United Nations Conference on Housing and Sustainable Development. Habitat 111: *New Urban Agenda*, Draft outcome document for adoption in Quito, October 2016.

UNFCCC. (2007). Synthesis and Assessment Report of the Greenhouse Gas Inventories Submitted in 2006. Note by Secretariat, FCCC/WEB/SAI/2006 28 June 2007

Van der Heijden, J. (2015). What 'Works' in Environmental Policy-Design? Lessons from Experiments in the Australian and Dutch Building Sectors. *Journal of Environmental Policy & Planning, 17*(1), 44–64.

Van Dooren, W., Bouckaert, G., & Halligan, J. (2008). *Performance Management in the Public Sector*. Hoboken: Routledge.

Van Thiel, S., & Leeuw, F. L. (2002). The Performance Paradox in the Public Sector. *Public Performance and Management Review, 25*(3), 267–281.

Walker, R. M. (2003). Evidence on the Management of Public Services Innovation. *Public Money & Management, 23*(2), 93–102.

Wheelan, C. J. (2011). *Introduction to Public Policy*. New York: WW Norton & Company.

Wheeler, S. M. (2008). State and Municipal Climate Change Plans: The First Generation. *Journal of the American Planning Association, 74*(4), 481–496.

World Bank. (2014). *A Wicked Problem: Controlling Global Climate Change*. http://web.worldbank.org/WBSITE/EXTERNAL/NEWS/0,,pagePK:34382~piPK:34439~theSitePK:4607,00.html

2

City Governments and Climate Change

Cities, What Are They Good For?

At the September 2015 climate summit convened by Pope Francis in Rome, the Mayor of Minneapolis, Betsy Hodges, claimed: 'The decisions we make in Minneapolis have an impact on things that happen in Ghana, or happen in Honduras' (Golden 2015). Hodges represents the views of many city mayors in taking a broad perspective and arguing that cities, because of their large populations and economic roles, make a contribution to global climate change. Along with 60 other mayors handpicked by the Vatican from South America, Africa, the United States, Europe, and Asia, Hodges signed a declaration warning world leaders preparing to meet in Paris that the December COP21 summit 'may be the last effective opportunity to negotiate agreements that keep human induced warming below 2 °C' (Pope Francis 2015). The mayors at the Pope's summit show the lengths to which some will go to try to appeal to international negotiations on climate change. There are, however, many mayors who do not take the same view as those signing the Vatican's declaration. For some, there are boundaries that can act like a wall around a medieval city in directing their focus and limiting their priorities. A 2014 survey of mayors

S. Jones, *Cities Responding to Climate Change*,
DOI 10.1007/978-3-319-64810-1_2

from the largest US cities revealed that more than 70% of Republican mayors believed their cities should not expend resources to mitigate the risks of climate change (Einstein et al. 2014). Some of the mayors who adopt this focused position argue against the reality of human contributions to climate change, others argue they can only focus on issues facing their city, as they are constrained by contextual factors such as politics, institutional arrangements, and resource limitations. The juxtaposition between the more focused and the broader perspectives of mayors is interesting in terms of what it reveals about how city leaders respond to issues, threats or otherwise, posed by climate change.

The scientific consensus conveyed through the Intergovernmental Panel on Climate Change (IPCC) concludes that climate change is a serious threat faced by all governments (Seto et al. 2014). In his speech to the UN Security Council in September 2014, President Obama argued that climate change posed a bigger threat to humans than terrorism, instability, and disease (Kenny and Cox 2015). Many city leaders align with this view and have adopted conclusions of IPCC scientists that risks to urban environments are increasing and threaten economic and environmental assets (IPCC 2014a; CDP 2014). There are thousands of city governments that have identified climate change as a policy problem and have taken some form of policy action. The actions of these cities provide an opportunity to examine fundamental questions of effectiveness and accountability in the implementation of climate policy. Unfortunately, evidence suggests that, where city governments are concerned, there continues to be a 'stubborn gap between the rhetoric and the reality of climate policy' (Betsill and Bulkeley 2007). Current IPCC assessments claim that 'there has been little systematic assessment regarding the overall extent to which cities are implementing mitigation policies and emissions reduction targets are being achieved, or emissions reduced' (Seto et al. 2014, 928). If cities are capable of climate action that 'surpasses anything countries can hope to do', then, in the context of the conclusions of the IPCC, it is of critical importance to examine the role of city governments in establishing effective climate policy implementation processes and procedures that measurably demonstrate success.

Climate Change and the City

Cities are regarded as both the 'victim and the culprit' of climate change (Bulkeley 2013, 9). Since 2007, the IPCC has argued that, as a result of anthropomorphic greenhouse gas (GHG) emissions leading to climate change, cities will be increasingly vulnerable to heat waves, extreme weather events, and rising sea levels. Many factors contribute to this increase, including reliance on fossil fuels, such as coal-fired power plants, dependence on private transport motor vehicles, inefficient use of energy in buildings, and the use of biomass for cooking and heating. It is the relationship between lifestyles and infrastructure patterns in highly urbanised areas that drives emission patterns and the emissions triggered by consumption in cities—commonly referred to as the carbon footprint. The carbon footprint of cities is largely determined by socio-economic factors, including growing income, education, and car ownership, as well as decreasing household size (Minx et al. 2013). In its 2014 report, the IPCC Working Group on Human Settlement argued that energy use and GHG emissions from human settlements are mainly from urban areas rather than rural areas, and the role of cities in global climate change has become increasingly important over time (Seto et al. 2014, 930).

The World Health Organization (WHO) and the IPCC both provide evidence to support the claims made by city leaders who argue that cities play a critical role in the challenges of climate change. The WHO recognises that some cities are making notable improvements, demonstrating that 'air quality can be improved by implementing policy measures such as banning the use of coal for heating in buildings, using renewable or cleaner fuels for electricity production, and improving efficiency of motor vehicle engines' (WHO 2014, 1). As an incentive for improvement to the current situation, the IPCC argues that there is considerable evidence to support the view that successful implementation of urban GHG mitigation strategies can produce 'co-benefits' of public savings, improved air quality, health benefits, and productivity increases (IPCC 2014a). Data from the World Bank (Table 2.1) shows a range of benefits experienced in some cities from implementing policies that mitigate and adapt to climate change. In many instances, city governments, and mayors in

Table 2.1 Co-benefits of mitigation and adaptation policies for city development

City	Action	Integrated value
Mexico City	Improvements to water supply infrastructure	Increases water supply Reduces vulnerability Increases access to basic services by the poor
Dar es Salaam	Coastal and marine conservation project	Sequesters carbon via mangroves Protects city from storm surges Healthy coastal ecosystem
Makati City	City-wide tree planting	Sequesters carbon dioxide Reduces pollution Provides food supply Reduces urban heat islands Recreational space
Bogota	Urban agriculture	Reduces need for fertilizers and pesticides Food supply Prevents settlement in high-risk areas
Lviv	Energy efficiency programme	Reduces energy consumption Buildings better able to cope with extreme temperatures

Source: Adapted from World Bank: Cities and Climate Change (2010)

particular, are keen to promote the economic benefits that will flow to their city if climate policies are implemented (Krause 2011). Whether city governments alone have the power, or even the will, to undertake the policy and regulatory action required is a point of some contention. Effective action can largely depend on contextual factors such as institutional arrangements, governance mechanisms, and financial resources that impact on city governments. What is reasonably clear from the evidence, however, is that cities are key focal points for anthropogenic activitythat will significantly contribute to and impact on a range of symptoms of climate change.

The co-benefits arguments identified in IPCC and World Bank reports have inspired city governments to establish plans designed to mitigate and adapt to the likely impacts of climate change. Mitigation-focused plans can include GHG reduction targets aimed at reducing emissions by reducing sprawl, improving energy efficiency, and promoting renewable energy. One of the most comprehensive of these plans was developed in

2007 by New York City under the leadership of Mayor Michael Bloomberg. *PlaNYC: A Greener, Greater New York* set a target of reducing emissions in the city by 30% by 2020 through a series of mitigation measures aimed at buildings and transportation. Bloomberg's plan included initiatives to help prepare New York for 'growth, an aging infrastructure and an increasingly precarious environment' (PlaNYC 2007, 4). The central focus of PlaNYC was to mitigate emissions in the city: 'All of PLANYC's strategies—from reducing the number of cars to building cleaner power plants to addressing the inefficiencies of our buildings—will help us to reduce emissions' (PlaNYC 2007, 13). In many cities, climate change plans also include a focus on adaptation measures such as new infrastructure to accommodate surges from increasingly intense storms and floods. For example, in 2011, the Mayor of London, Boris Johnson, released the *Managing Risks and Increasing Resilience: The Mayor's Climate Adaptation Strategy* (GLA 2011). The strategy draws attention to flooding, drought, and overheating as consequences of climate change for which London needs to prepare. As further evidence for the co-benefits argument, Mayor Johnson stressed that while the adaptation strategy would improve the ability of the city to cope with projected climate shifts, he recognised the economic benefits of the actions taken: 'Inherent in this goal is the drive to attract significant new investment into London, thereby securing tens of thousands of jobs through the delivery of vital infrastructure, goods and services, as well as developing world-class exportable skills' (GLA 2011, 5).

International organisations such as the Local Governments for Sustainability (ICLEI) have been promoting the significant contribution city governments can make in responding to climate change. The ICLEI provides comprehensive guidelines for establishing plans and strategies which, like the *Urban Low Emission Development Strategy*, developed in conjunction with the UN, include Web-based case study examples to promote policy action among decision-makers (UN-Habitat n.d.). In a similar attempt to encourage necessary participation by city governments, the UN-Habitat (2015) announced the release of *Guiding Principles for City Climate Action Planning* to coincide with the COP21 conference in Paris. The Organisation for Economic Co-operation and Development (OECD) also argues that cities can be laboratories for innovative

Table 2.2 Leading actors in climate change experiments

	Leading actor	Local government	Other government
Where	Africa	29	6
	Asia	86	13
	Europe	112	11
	North America	102	11
	Oceania	15	1
	South and Central America	69	24
When	Pre Kyoto agreement	20	4
	Pre Kyoto ratification	65	8
	Post Kyoto	328	54
What	Adaptation	46	19
	Built environment	101	13
	Carbon sequestering	16	8
	Transport	96	140
	Urban form	27	2
	Waste management	18	0
	Water management	10	10
	Energy	99	13
Total		413	66

Source: Adapted from Broto and Bulkeley (2013, 98)

approaches to climate policy (OECD 2010). Table 2.2 shows results of recent quantitative research of 100 cities across the world which confirm that local governments have taken a prominent role in urban climate change experiments, including intervention in energy production and urban infrastructure (Broto and Bulkeley 2013, 98). 'The sheer magnitude of policy initiatives at the local level, their diversity, and the experimental and practical nature of many local projects, are bound to bring forward genuinely new ideas and solutions that in the end can have an impact on a larger scale' (Gustavsson et al. 2009, 72). A central assumption of local climate change protagonists is that initiatives introduced by city governments can be scaled up to regional and national levels by acting as pilot studies that can demonstrate what can be effective in mitigation and adaptation. However, research also shows that national and subnational governments tend to place limitations on city governments and restrict their capacity to implement effective climate change policy (Jones 2012; Bulkeley and Betsill 2005). As a result, city governments argue that, despite their potential, they have been largely ignored in national and international debates on climate policy solutions.

Multiple and Conflicting Objectives

Assessments of responses to climate change must consider the potential objectives that can influence the approaches taken. The actual intentions behind the public statements in responding to climate change will have consequences for practitioners responsible for implementing the policy initiatives of the city government. The policy literature emphasises the importance of goal clarity to implementing public policy. Senior managers provide the link between practitioners and politicians. It is their role to ensure that the intent of the policy is clear through performance systems that impact on the goals of both organisational units and individual staff members in their performance assessments. Arguments that dispute the real intent of the city governments claiming to be responding to climate change provide interesting perspectives on the motivations of city leaders. For some researchers, sustainability policies are the latest form of boosterism (McCann 2013). The dilemma this presents for climate policy is that it fails to provide clarity in objectives for decision-makers and policy practitioners within city governments. This failure will also impact on a performance-based approach. The policy literature outlines the significance of a lack of goal clarity, which can be serious on a number of levels for both the organisation itself and for individual practitioners. Cities are presenting 'clean' and 'green' images as part of an attempt to establish local economies that support innovation and creative industries. By pursuing the 'green' agenda, city governments are attempting to appeal to the 'creative class', described by Richard Florida (2014) as highly qualified individuals with skills to help catalyse the transformation of old economies to the new digital and creative environments that help establish opportunities for business investments. Many cities have been adopting entrepreneurial city marketing approaches that utilise sustainability and climate change in a competitive way that is 'fundamentally about comparing, contrasting, and ranking cities for specific purposes' (McCann 2013, 7). For example, in Rotterdam, the Mayor (Ahmed Aboutaleb) is enthusiastic in showing the international media new waterfront developments in formerly poor, industrial neighbourhoods, which he argues show how urban renewal dovetails with strategies to mitigate the effects of climate change ((Kimmelman 2017). A competitive attitude by city government leaders is focused specifically on potential business investors, the coveted creative workers, granting agencies, and tourists. Boosterism activities are relevant to discussions about climate policy because they are often designed to direct attention to certain places and policy initiatives, showing them to be exemplary and therefore regarded as benchmarks for comparison and objects of competition and emulation.

Boosterism activities are arguably part of a much larger phenomenon that is impacting on the democratic values that underpin city government. Writers such as Peck, Brenner, and Swyngedouw see cities as the focal point of neoliberal transformations that focus on the primacy of markets and capital at the expense of citizens and democracy. They see a*neoliberal urbanism*where city governments have been forced to manage responsibilities transferred from other levels of government, commonly without the requisite capacity to raise revenue. This cost-shifting by higher levels of government has helped to create a necessity for cities to focus on urban policy regimes that promote economic opportunities through neoliberal approaches such as 'de-regulation, privatisation, liberalisation and enhanced fiscal austerity' (Peck et al. 2009, 58). Climate policies that feature place marketing, tax abatements, public–private partnerships, and property development can be seen to be part of this agenda. The overarching goal of these approaches by city governments is to 'mobilise city space as an arena for both market oriented economic growth and for elite consumption practices' (ibid.). The linkages between these objectives and the climate policies of city governments need to be explored, as they may reflect, what can be described as, cynical approaches to climate policy that are known to be ineffective in reducing GHG emissions but are actually intended for marketing and promotional purposes.

Networks of City Governments

Many mayors argue that the comparative lack of climate action from higher levels of government (Table 2.2) has catalysed establishing networks and coalitions to strengthen their advocacy role and promote their *bona fides* in national and international climate negotiations. At the UN Climate Change Conference in Copenhagen in 2009, 80 city mayors held their own *Climate Summit for Mayors* to push for greater recognition of the role of cities in negotiations over climate change policies. The communiqué from the Mayors' summit asked a key question of other levels of government:

> Why do your negotiations exclude the government actors who have demonstrated the political will and practical ability to start solving this global problem? The stakes are too high to be constrained by an antiquated process that mutes the voices of millions. Bring us in as partners, give us a seat at the table and let us share in this responsibility. (ICLEI 2009)

This collective plea by mayors from some of the world's largest cities represents concern over the lack of collaboration between local and national levels in implementing climate change policies. A more localised example is the Southeast Florida Regional Climate Change Compact established in 2007, consisting of 27 local governments, including the City of Miami Beach. The Mayor of Miami Beach, Philip Levine, was elected in 2013 with a commitment to deal with the ongoing and increasing sea level rise causing flooding throughout the city. Along with other members of the Regional Compact, the City of Miami is undertaking a Regional Climate Action Plan (SEF 2012). In a clear example of extreme state resistance, despite the efforts of the city governments, Florida's Governor, Rick Scott, instructed state government workers not to discuss climate change, or even use the term (Kolbert 2015). While there is little acknowledgement of the local government action by Florida's governor, President Obama described the Regional Compact as a model for other local governments, 'not just for the country, but for the world' (Obama 2014).

Regional, national, and international city networks are seen as a means of raising profiles and achieving climate policy objectives by many like-minded city governments (Young 2007). Since the early 1990s, many cities in North America and Europe have been playing a role in international networks of localities attempting to make progress in reducing GHG emissions (Osofsky and Levit 2007; Giest and Howlett 2013). A key underlying assumption is that properly managed collaborative efforts can develop solutions most effective for a local context, while at the same time offering access to best practice options through connections with knowledgeable actors, resources, and government agencies (Giest and Howlett 2013).

Such networks have been established to further the interests, and improve the effectiveness, of member governments. Participating in networks provides governments with access to opportunities, and allows the municipality itself to take a lead role in climate policy. Leaders of city governments argue that such cooperation can open opportunities of international recognition, in order to stand out as a pioneer city welcoming innovative ideas, combining local economic development with reduction of GHG emissions (Gustavsson et al. 2009). Some of the benefits experienced by members of the *Cities for Climate Protection Campaign* (CCP) network include the production and dissemination of technical information about local contributions to climate change, measures that can be taken locally to address the problem, and the potential co-benefits (Betsill and Bulkeley 2004). Other benefits include implementation advice derived from the experiences of member cities. The ICLEI provides advice and guidelines for city government members on implementation approaches to adopt. The ICLEI's CCP initiative milestones include establishing an emissions inventory, setting GHG reduction goals, developing a plan to meet goals, implementing the plan, and monitoring and evaluating the results. These also include extensive research and comparison of member governments that have undertaken measures to implement climate policy action.

European city governments have been effective in working as coalitions to influence climate policies by higher levels of government. Cities in France, Germany, and Norway, through *transnational municipal networks* (TMNs), have attempted to exert an influence in two key ways: First, the administration and activities of city governments should be more sustainable and climate neutral. Second, cities should be attempting to influence behaviour change by citizens and businesses towards more sustainable practices (Giest and Howlett 2013). In this respect, European cities have been more effective at implementing climate change policies than their North American counterparts. The small Swedish city of Vaxjo is widely regarded as a leading example of how city governments can promote change. Famous since 1996 for being the first city to set the goal of being fossil fuel free, Vaxjo has been working together with a coalition of local industry and other levels of government to successfully reduce GHG emissions by 30% by promoting non-fossil fuels such as solar, encouraging alternative forms of transport to promote health and reduce emissions, and prohibiting urban sprawl (Gustavsson et al. 2009).

Whatever their tangible effects, networks can be regarded as an expression of 'policy proximity', that is, coalitions of geographically dispersed actors drawn together by a will to act jointly in order to mitigate and adapt to climate change (Gustavsson et al. 2009). The participants in these networks do not hesitate to cross levels and scales when undertaking this work. In the view of some commentators, these networks are becoming 'true creators of global environmental governance' (ibid., 63). However, while networks may strengthen the participants' ability to attract investments from the private sector and from public funding to bring about sustainable development, they are also subject to provisions from national and regional governments, which might hamper their benefits. Evidence from Europe suggests that city networks that target a specific region and are supported by other levels of government have the most benefits for climate change. However, there is also some evidence that there seems to be considerable overlap between some of the networks and many pay lip service to climate change actions that are not translated into action (Betsill and Bulkeley 2006). More recently, networks have been accused of 'orchestrating' responses by member city governments by steering them towards particular actions and objectives. Current research focuses attention on efforts to promote coordinated efforts and activities between cities by the assumption that orchestration represents both a conscious strategy that incorporates and coordinates the actions, decisions, and metrics being used to evaluate cities and city networks in urban climate governance, and an 'orientation that reflects the dissemination and diffusion of experimental norms, metrics and best practices' (Gordon and Johnson 2017, 708). There is some evidence that there are subtle forms of power of networks that underlie seemingly benign efforts to coordinate actions and metrics at multiple scales. Other researchers argue that more 'strategic and less opportunistic networking is needed' (Acuto et al. 2017, 14). These authors suggest that there might be too many city networks, contributing to a confusing global landscape and raising questions of possible natural selection among networks in a context where resources are constrained (ibid.).

A central feature of city climate networks is to build regular channels of communication among public and private actors in the environmental field, and that they contribute to a learning effect among cities. Networks can impact member cities by accelerating learning processes, which they

Table 2.3 Different types of transnational municipal networks

	Extent	
	Global	Regional
Small number of members	Global high-profile networks: World Mayors Council on Climate Change (WMCCC) C40 Cities Clinton Climate Initiative (CCI) Sustainable Cities International Network (SCI)	Regional high-profile networks: Eurocities Cities Development Initiative Asian Cities Climate Change Resilience Network (ACCCRN)
Large number of members	Global in-depth networks: Cities for Climate Protection Campaign (CCP) Local Governments for Sustainability (ICLEI) United Cities and Local Governments (UCLG) The Climate Group	Regional in-depth networks: Climate Alliance Energy Cities Covenant of Mayors Mayors Climate Protection Center

Source: Adapted from Giest and Howlett (2013, 350)

'achieve by increasing the availability of information, by multiplying contacts among members and by enhancing their capacities to act locally on climate change' (Hakelberg 2011, 73). Policy learning contributes to identifying key factors that contribute to the successful implementation of complex policies such as those dealing with climate change (Table 2.3).

Implementing Climate Policy

Policy implementation is 'what develops between the establishment of an apparent intention on the part of government to do something, or stop doing something, and the ultimate impact in the world of action' (O'Toole 2000, 266). Issues that impact on the implementation process are particularly important to understand for climate policy, as it is the crucial stage where policy decisions are translated into action; it is the process whereby programmes or policies are carried out. This is an important issue for both individual city governments and network coalitions because success will be critical in promoting the co-benefits of pursuing climate change initiatives. Public administration and public sector management research plays an important role in this area, as there is considerable research on policy formation, issues of governance, and policy implementation (Howlett and Ramesh 2003; Taylor and Balloch 2005). Policy implementation research has been directed at a broad cross section of government policy activity. Implementation of climate change policy is an emerging area of research and provides a number of challenges common to other 'wicked' policy problems. These problems go beyond the capacity of any one organisation to understand and respond to, and where there is often disagreement about both the causes of the problems and the best way to respond to them (APSC 2007). Wicked problems pose challenges for policy practitioners, as they require multidisciplinary approaches across multiple agencies and across different levels of government. As an example, the Australian government reflects the view of many governments when it regards climate change in this way:

> Climate change is a pressing and highly complex policy issue involving multiple causal factors and high levels of disagreement about the nature of the problem and the best way to tackle it. The motivation and behaviour of individuals is a key part of the solution as is the involvement of all levels of government and a wide range of non-government organisations. (APSC 2007, 1)

When considering climate policy, researchers are recognising serious problems in terms of policy design as well as implementation styles (Rykka et al. 2013). Climate change as a policy problem transcends structural and functional sectorial boundaries, administrative organisational levels, and distinctions between political actors and traditional policy. Public administration literature reveals that effective and long-term management of climate policy is typically hindered by traditional hierarchical and departmentalised methods of governing, complex interdependencies, and difficulties of definition and identification, especially in reaching consensus on solutions (ibid.). These are all representations of the challenges presented by the implementation of climate policy. The examination of implementation issues is under resurgence, as policy practitioners have traditionally found themselves 'enmeshed in the vexing challenges of converting policy intent into efficacious action' (O'Toole 2000, 265).

Important issues have emerged in the literature regarding the factors that impact on policy implementation that will have relevance to climate policy. Questions have been raised about the efficacy and clarity of policy commitments since Pressman and Wildavsky's (1973) seminal research on the topic revealed the complexities and contradictions of implementation. Their research referred to the challenges that emerge in 'chains of interaction' required to implement policy: their concept of an 'implementation deficit' being likely the longer the chain. The increase in the layers of management between policy-makers and practitioners responsible for delivery, particularly in multilevel systems, can contribute to the implementation deficit. This is a valuable lesson for the implementation of climate policy. The more levels of government, the more departments and agencies, and the more hierarchical management arrangements between organisational units involved in policy implementation, the more opportunities there can be for implementation to divert policy outcomes from the original intentions. Current research suggests that managerial competence and governance skills may help reduce variation in implementation results and may even compensate for the lack of goal clarity of the original policy (Hupe 2011). Analysis of complexities in implementing climate policy warrants ongoing investigation, as they have the potential to alter the outcomes governments are seeking.

Within city governments, climate policy requires attention and action from different units, for example, infrastructure, planning, environment, economic development, and finance. Coordination, or mainstreaming, across these units will be critical elements of the implementation process put in place at both the political and managerial level. Current research concludes that 'we know very little about the way sustainability and climate policy efforts are (or should be) organized in cities, particularly whether local governments have established the administrative arrangements and institutional changes necessary to facilitate the success of environmental sustainability efforts' (Krause et al. 2016, 115). To help illustrate this point, research on Canadian city governments reveals that there are barriers within city government administrations that can inhibit action on climate change (Burch 2009). Barriers can include an organisational culture of combativeness between functional areas. The same research suggests that a culture of collaboration or innovation can act as an enabler of action that is critical to the implementation of climate policy (ibid.). The City of Stockholm has attempted to deal with such issues through a level of integration in budgetary and planning processes. For the implementation of climate policy measures within its administration, the city government claims that sustainability principles are integrated across all strategies and city planning. According to the city council, the integrated budget and administrative systems guarantees that environmental aspects are considered in budgets, operational planning, reporting, and monitoring (City of Stockholm 2015). In terms of climate policy, current research points to rather few traditional implementation studies, that is, studies investigating processes of implementation and the implementation of specific programmes or policy measures (Rykka et al. 2013, 109, 125). Further studies are needed that deal with how the process of implementation takes place and strive to identify existing problems and solutions related to successful implementation.

Important implementation issues also emerge from the work of Lipsky (1980) on the impact of 'street-level bureaucrats' as the individuals with direct interaction with citizens and with wide discretion over the delivery of services. Policy implementation roles of street-level bureaucrats are based on relatively high degrees of discretion and autonomy from organisational authority. This is particularly relevant for city governments as the level of government closest to citizens. Interaction between city bureaucrats and citizens will be critical for climate policy as the focus increasingly turns to behaviour change for improved sustainability and commitment to mitigation and adaptation initiatives. Lipsky observes that public employees who interact with citizens can manage their difficult jobs by developing routines of practice and psychologically simplifying their environments in ways that strongly influence outcomes. Street-level bureaucrats develop coping behaviours in complex and challenging environments that often reflect compromises between goals of enacted policy and the needs of the bureaucrats (Lipsky 2010, xvii). Other researchers support this conclusion and argue that street-level bureaucrats behave in ways that are 'unsanctioned, sometimes even contradicting official policy, because the structure of their jobs makes it impossible to fully achieve the expectations of their work' (Hupe and Hill 2007, 279). Public administration research continues to build on Lipsky's original observations, with recognition that the implementation of policy decisions takes place within a 'nested sequences of decisions—about structure, financing, and about the management of outputs—for which different actors may be accountable, perhaps in different ways' (Hill 2005, 277–278). This research raises the importance of the institutional context and the multiple accountabilities of practitioners—bottom up as well as top down, and also sideways (to their profession). This context has implications for climate policy, as how policy practitioners act can be shaped by institutional, political, and practical factors (Hupe and Hill 2007, 296). As an example, analysis of the implementation of the New York City Government's PlaNYC shows that cross-agency initiatives require processes that reduce resistance from individual employees based on anticipated blame for failure (Jones 2015). Building trust based on linking the competent management of people to improvements against organisational goals can improve the likelihood for change.

The experience across OECD countries shows that performance targets imposed on top of existing systems become an additional layer of control that can promote staff resistance in already overburdened systems (OECD 2008). Effective performance management systems are promoted as offering solutions to the implementation issues raised here. Protagonists argue that a performance-based approach to management can be established to monitor progress, better allocate resources, make decisions about strategy, reengineer processes, motivate workers, and usher in a new era of accountability (Moynihan 2009). Performance management is important to the implementation of climate policy, as it has the potential to provide answers to important implementation question about linkages between government objectives, administrative and management processes designed to achieve the objectives, and the measurement and reporting of results. Performance management is an 'act' of management where actions begin by observing the current state of performance, committing to a more favourable level of performance, and taking action to achieve a targeted level (Ammons 2015).

The Measurement and Management of Performance

Performance management has the potential to enable governments to quantify promises, and measure and report their actions in ways that allow citizens, managers, and politicians to make meaningful decisions about increasingly complex government activities (OECD 2009). New Public Management principles promoted by governments since the late 1980s offered a broad rationale for public sector performance measurement and reporting. Reforms based on these principles have encompassed all levels of government and featured a focus on results and on measuring and publicly reporting performance results as a principal way of increasing accountability. As a direct link to policy implementation, embedded within performance measurement is the assumption that measurement and reporting can also drive improvements in the efficiency and effectiveness of government organisations (Dubnick 2005). The promise is that measurement—the quantitative representation of the quality or quantity of inputs, outputs, and/or outcomes (Sterck and Scheers 2006)—can offer valuable data for policy-makers by assisting with comparison, prediction, and evaluation. Measuring performance involves the systematic collection of data by observing and registering performance-related issues for a performance-related purpose (Bouckaert and Halligan 2008, 26). Performance measurement is a tool primarily used for accountability and reporting purposes—'good reporting allows for transparency, verification, and replication over time' (Ibrahim et al. 2012, 223). Reporting is an integral part of a performance measurement process, as it provides valuable information that contributes to holding managers to account for the work of their organisational units. In climate policy, this can be particularly important when reporting results helps to legitimise the decisions taken and helps citizens understand the actions being taken.

Effective performance management systems for mitigation and adaptation policies will require a foundation of good measurement systems, with metrics and agreed accounting methodologies that develop universal protocols for establishing baseline data. However, this foundation is problematic in cities, as they are 'porous' open systems, and physical and political boundaries, however defined, will influence the estimates of emissions (IPCC 2014a, 936), see Table 2.4, has identified what Kennedy and his colleagues argue are the majority of currently known sources of emissions, there may be more but only further research will tell (Kennedy et al. 2010). Current measurement protocols established to develop inventories for measuring emissions in cities do not cover all available sources. The popular method developed by the ICLEI, while a reasonable starting point, does not identify all sources of emissions. A substantial concern with measurement researchers is the failure to consider consumption-based emissions. With cities generating approximately 70% of GDP and with approximately 50% of the world

Table 2.4 Comparison of measures for attributing GHG emission to cities considering spatial boundaries and the life-cycle perspective

WRI/WBCSD definition	Spatial boundary	Life-cycle perspective	Components
Scope 3	Out-of-boundary energy use (and further out-of-boundary emissions not included in Scope 2)	Production chain emissions	Embodied emissions from food and materials consumed in cities
			Emissions upstream of electric power plants
			Upstream emissions from fossil fuel use
		Single-process emissions	Combustion of aviation and marine fuels
			Out-of-boundary waste (landfill) emissions
			Out-of-boundary district heating emissions
Scope 2	In-boundary electricity use		Out-of-boundary electricity emissions at power plants
Scope 1	In-boundary emissions		In-boundary fossil fuel combustion
			In-boundary waste (landfill) emissions
			In-boundary industrial processes and product use
			In-boundary agriculture, forestry, and other land use

Source: Kennedy et al. (2010).
WRI, World Resources Institute; WBCSD, World Business Council for Sustainable Development

population not counting emissions from what they consume has been a source of ongoing debate. Governments have been reluctant to consider these emissions as part of their measurement protocol. Current estimates suggest that emissions from the production of internationally traded products are as high as 25%; however, because they are manufactured elsewhere, the emissions are difficult, if not impossible, to measure.

Researchers in the United Kingdom have examined the issues surrounding consumption-based emissions at the national level and their results reflect those for other developed nations (Fan et al. 2016; Barrett et al. 2013). Their argument is that complexity is one of the significant inhibitors to making accurate calculations. They argue that current political leaders are reluctant to even consider policy options, political responses, and institutional and governance issues associated with these emissions (Barrett et al. 2013). There is not a great deal of research on this issue at the city level. Chinese researchers have examined the consumption of carbon dioxide (CO_2) emissions from 13 Chinese cities and found significant differences between consumption- and production-based emissions. In these cities, 70% of consumption-based emissions are imported from other regions. As cities become more advanced and socio-economic development increases wealth, consumption-based emissions exceed production-based emissions (Mi et al. 2016). In terms of city government efforts to respond to climate change, this is a very important finding for developed countries. It suggests that unless city governments tackle consumption-based emissions, they are not really dealing with the major impact their communities are having on global emissions. One of the few studies examining more closely the emissions from consumption at the city level has been undertaken in Kyoto. Researchers found that it would be possible to reduce emissions by 80% by 2050 by following some reasonably straightforward ideas that focused on energy saving through electrification of all transportation, the promotion of wood utilisation for housing and household energy saving, the introduction of renewable energies, and efficient energy recovery from wastes. Shifting the final consumption mode into the low CO_2 emission mode can make a significant contribution in reducing emissions according to conditions such as household composition, economic conditions, and climate (Shigeto et al. 2012).

Despite these difficulties, many city governments have been establishing performance measurement schemes that align with their climate change action plans. Some cities have established reporting systems that reflect an attempt to move beyond measurement to performance management frameworks intended to improve the implementation of such plans and the achievement of outcomes. As an example, the City of Philadelphia has created *PhillyStat* as its performance management and tracking tool that involves a twofold review process focusing on both operational and outcome levels (Nam 2014).

According to the City of Philadelphia, the main goals of the *PhillyStat* programme apply across all operational units of the city government. The goals include:

- To drive better results and services, by maintaining a focus on problem-solving and process improvement;
- To bring the right information to the table, by integrating operational, financial, and other relevant information into performance sessions;
- To set meaningful measures and targets in a collaborative fashion;
- To manage consistent priorities, by placing emphasis on the City's top priorities;
- To better communicate with stakeholders. (City of Philadelphia 2015)

PhillyStat is based on the 'Stat' model that has been adopted by many governments and reflects a performance management system that incorporates both measurement and cross-agency collaboration. Each city is able to develop a Stat system with unique properties to suit its idiosyncrasies. Figure 2.1 highlights the performance goals and objectives of the Sustainability Office of the City of Philadelphia. Staff from the office are required to report on their progress, both on operational-level performance and on an outcome-level performance review, highlighting successes and outlining plans for improvements where necessary (Nam 2014).

The experiences of cities with Stat-based systems such as Philadelphia have contributed to identifying a range of challenges associated with developing indicators needed for performance management systems. Current research provides an insight into the challenges faced by governments that are typical to developing indicators for measuring performance. Table 2.5 highlights the key challenges based on international

Fig. 2.1 PhillyStat sustainability report 2015
Source: City of Philadelphia (2015)

Table 2.5 Challenges in developing indicators

1. It is challenging to determine what to measure.
2. Assessment of costs involved in measuring multiple variables can be complex.
3. Attempts at influence measurement through political means are common.
4. It is difficult to ensure whether measurement data is reliable and can be replicated in different cities.
5. Political commitment and funding is inconsistent.
6. Some variables defy measurement—for example, trust and confidence.
7. Political boundaries can vary over time.
8. Data generated by non-city agencies may be difficult to access.
9. Regular updating information is often uncommon.

Source: Adapted from Hoornweg and Friere (2013) and Hoornweg (2015)

assessments (Hoornweg and Friere 2013; Hoornweg 2015). Research shows that climate action plans are susceptible to a range of common issues that may work against the overall outcome of tackling climate change (Seto et al. 2014). Such issues include being largely focused on energy efficiency rather than on broader land-use planning strategies and cross-sectoral measures to reduce urban sprawl and promote transit-oriented development. The majority of these targets have been established in developed countries and reflect neither their mitigation potential nor implementation. Few targets have been established for non-developing country cities, and it is in these places that a reliable city-level GHG emissions inventory may not exist (IPCC 2014b, 5).

International standards for GHG inventories and measurement schemes are being developed by researchers and by networks of city governments (Baur et al. 2015; C40 Cities 2015). One of the leading examples is the C40, which has links to the Clinton Climate Initiative (CCI), the ICLEI, and the Climate Group. The C40 was established in 2005 with 13 affiliate city governments from the United States, Europe, and China (Schreurs 2008; Bulkeley 2010; Roman 2010). The network continues to expand, and in 2015 claimed to have more than 80 city members 'representing 600+ million people and one quarter of the global economy' (C40 2015). The C40 claims to have developed the most comprehensive measurement methodologies in conjunction with the ICLEI and the World Resources Institute (WRI 2015). The *Global Protocol for Community-Scale Greenhouse Gas Emissions Inventories* (GPC) has been developed over a number of years as the first common standard to measure and report city emissions. The WRI argues that the GPC 'will help cities see what climate strategies are working, better target their resources, and hold themselves accountable for results. The more cities take part in the Compact and adopt the GPC, the greater impact it will have' (WRI 2015). A pilot study involving 35 cities from developed and developing nations has been underway since 2016. Preliminary trials are providing information in the development stages of the GPC (Fig. 2.2).

Despite the efforts of many individual cities and international networks, the experience continues to show that measurement and reporting are challenging exercises. Table 2.6 provides a summary of research by the IPCC that reveals that, while current measurement and accounting methods have been useful, key limitations continue to restrict comparability across the systems (Seto et al. 2014, 935–939).

City governments are undertaking climate action plans on a local and international scale, but research shows that their aggregate impact on urban emissions is uncertain (Seto et al. 2014). Local governments, individually and in coalitions, possess unique opportunities to engage in urban mitigation activities. Local mitigation policies continue to grow in

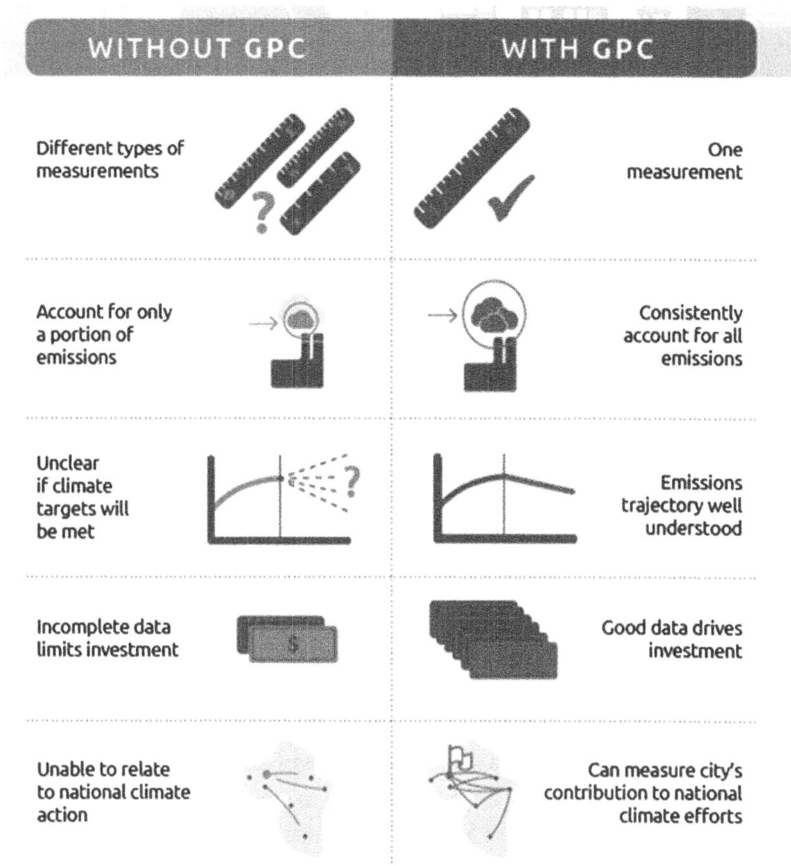

Fig. 2.2 C40: Features of the Global Protocol for Community-Scale Greenhouse Gas Emissions Inventories
Source: WRI (2015)

number and complexity. So far, experience in the development of performance schemes aimed at supporting the implementation of local climate policies has been that limitations emerge when establishing the protocols that are the foundations of an effective performance management system. Without improvements to these performance frameworks, there is a risk that the implementation of climate policy cannot be accurately measured, raising questions about the effectiveness and success of their original objectives. A central argument of this book is that city governments

Table 2.6 Limitations of current GHG measurement schemes

1. Publications are restricted to the analysis of energy consumption and GHG emissions from a limited set of emissions estimates. New estimates do not emerge at the same pace.
2. Available evidence is particularly scarce for medium and small cities, as well as for rural settlements.
3. Most studies focus on developed countries, with limited evidence from a few large cities from developing countries, for example, limited emissions estimates from China.
4. Emissions estimates are primarily from energy-related carbon dioxide emissions, not all GHG emissions.
5. While there is a considerable amount of evidence for territorial emissions, studies with a wider scope of emission components are growing but remain limited.
6. Due to marked differences in accounting methods, scope of covered sectors, sector definition, GHG emissions covered, and data sources used, there is a restricted capacity to compare across studies.

Adapted from Seto et al. (2014)

need to move from a performance measurement focus to a performance management-based approach. Measurement does not automatically lead to performance improvement. Research on performance schemes in local governments reveals that only those engaged in the act of performance management can point to changes in the organisation and improvements in service delivery made in response to performance information (Ammons 2015). By adopting a performance management-based approach, cities can move from collecting and reporting measures to a management approach that can improve implementation, with key features of analysing, discussing, and improving. This transformation will be an important consideration in the case study cities of Copenhagen, Stockholm, and Tokyo. Through these case study examples, we will become more aware of the methodological challenges faced by city governments attempting to measurably demonstrate their success through evidence of management and policy change contributing to sustainable climate outcomes.

In the next chapter, the book will focus on the climate policy initiatives of city governments.

References

Acuto, M., Morissette, M., & Tsouros, A. (2017). City Diplomacy: Towards More Strategic Networking? Learning with WHO Healthy Cities. *Global Policy, 8*(1), 14–22.

Ammons, D. (2015). Getting Real About Performance Management: Using Performance Information to Improve Services is Key. *Public Management*. December. ICMA Publications, PM Magazine. http://icma.org/en/Article/106489/Getting_Real_About_Performance_Management

APSC (Australian Public Service Commission). (2007). *Tackling Wicked Problems: A Public Policy Perspective*. Canberra: Australian Government.

Barrett, J., Peters, G., Wiedmann, T., Scott, K., Lenzen, M., Roelich, K., & Le Quéré, C. (2013). Consumption-Based GHG Emission Accounting: A UK Case Study. *Climate Policy, 13*(4), 451–470.

Baur, A. H., Lauf, S., Foerster, M., & Kleinschmit, B. (2015). Estimating Greenhouse Gas Emissions of European Cities—Modeling Emissions with Only One Spatial and One Socioeconomic Variable. *Science of the Total Environment, 520*, 49–58.

Betsill, M. M., & Bulkeley, H. (2004). Transnational Networks and Global Environmental Governance: The Cities for Climate Protection Program. *International Studies Quarterly, 48*(2), 471–493.

Betsill, M., & Bulkeley, H. (2006). Cities and the Multilevel Governance of Global Climate Change. *Global Governance, 12*, 141–159.

Betsill, M., & Bulkeley, H. (2007). Looking Back and Thinking Ahead: A Decade of Cities and Climate Change Research. *Local Environment, 12*(5), 447–456.

Bouckaert, G., & Halligan, J. (2008). *Managing Performance: International Comparisons*. London: Routledge.

Broto, V. C., & Bulkeley, H. (2013). A Survey of Urban Climate Change Experiments in 100 Cities. *Global Environmental Change, 23*(1), 92–102.

Bulkeley, H. (2010). Cities and the Governing of Climate Change. *Annual Review of Environment and Resources, 35*, 229–253.

Bulkeley, H. (2013). *Cities and Climate Change*. London: Routledge.

Bulkeley, H., & Betsill, M. (2005). Rethinking Sustainable Cities: Multilevel Governance and the 'Urban' Politics of Climate Change. *Environmental Politics, 14*(1), 42–63. DOI 10.1080/0964401042000310178.

Burch, S. (2009). In Pursuit of Resilient, Low Carbon Communities: An Examination of Barriers to Action in Three Canadian Cities. *Energy Policy, 38*(12), 7575–7585.

C40. (2015). *Cities 100: 100 Solutions for Climate Action in Cities*. C40 and Sustainia. http://www.sustainia.me/cities/

CDP (Carbon Disclosure Project). (2014). Protecting Our Capital: How Climate Adaptation in Cities Creates a Resilient Place for Business. http://icma.org/en/icma/knowledge_network/documents/kn/Document/306514/Protecting_Our_Capital

City of Philadelphia. (2015). Greenworks Philadelphia. http://www.phila.gov/mdo/phillystat/Pages/default.aspx

City of Stockholm. (2015). *Stockholm Action Plan for Climate and Energy 2012–2015*. Stockholm.

Dubnick, M. (2005). Accountability and the Promise of Performance: In Search of the Mechanisms. *Public Performance and Management Review, 28*(3), 376–417.

Einstein, K., Glick, D., & Lusk, K. (2014). *Mayoral Policy Making: Results from the 21st Century Mayor's Leadership Survey*. Boston: Boston University Initiatives on Cities.

Fan, J. L., Hou, Y. B., Wang, Q., Wang, C., & Wei, Y. M. (2016). Exploring the Characteristics of Production-Based and Consumption-Based Carbon Emissions of Major Economies: A Multiple-Dimension Comparison. *Applied Energy, 184*, 790–799.

Florida, R. (2014). *The Rise of the Creative Class—Revisited: Revised and Expanded*. New York: Basic Books.

Giest, S., & Howlett, M. (2013). Comparative Climate Change Governance: Lessons from European Transnational Municipal Network Management Efforts. *Environmental Policy and Governance, 23*(6), 341–353.

GLA (Greater London Authority). (2011). *Managing Risks and Increasing Resilience: The Mayor's Climate Change Adaptation Strategy*. London: GLA.

Golden, E. (2015). Mpls. Mayor: Debate Over Climate Change Not an Issue at Pope's Summit, *Star Tribune*, July 22.

Gordon, D., & Johnson, C. (2017). The Orchestration of Global Urban Climate Governance: Conducting Power in the Post-Paris Climate Regime. *Environmental Politics, 26*(4), 694–714.

Gustavsson, E., Elander, I., & Lundmark, M. (2009). Multilevel Governance, Networking Cities, and the Geography of Climate-Change Mitigation: Two Swedish Examples. *Environment and Planning C: Government & Policy, 27*(1), 59.

Hakelberg L. (2011). Governing Climate Change by Diffusion. Transnational Municipal Networks as Catalysts of Policy Spread, Forschungszentrum fuer Umweltpolitik, Freie Universitaet Berlin, FFU-Report 08-2011.

Hill, M. (2005). *The Public Policy Process*. Harlow: Pearson.

Hoornweg, D. (2015). *A Cities Approach to Sustainability*. Unpublished Ph.D. thesis, University of Toronto. Department of Civil Engineering.

Hoornweg, D., & Freire, M. (2013). Building Sustainability in an Urbanizing World: A Partnership Report. Urban Development Series Knowledge Papers No. 17. World Bank, Washington, DC: © World Bank. https://openknowledge.worldbank.org/handle/10986/18665

Howlett, M., & Ramesh, M. (2003). *Studying Public Policy: Policy Cycles and Policy Subsystems*. Oxford: Oxford University Press.

Hupe, P. L. (2011). The Thesis of Incongruent Implementation: Revisiting Pressman and Wildavsky. *Public Policy and Administration, 26*(1), 63–80.

Hupe, P., & Hill, M. (2007). Street-Level Bureaucracy and Public Accountability. *Public Administration, 85*(2), 279–299.

Ibrahim, N., Sugar, L., Hoornweg, D., & Kennedy, C. (2012). Greenhouse Gas Emissions from Cities: Comparison of International Inventory Frameworks. *Local Environment, 17*(2), 223–241.

ICLEI (International Council for Local Environmental Initiatives). (2009). ICLEI Statement of Mayors and Local Governments on Failure of International Community to Act on Climate Change. http://www.icleiusa.org/newsevents/press-room/press-releases/iclei-statement-of-mayors-and-local-governments-on-failure-of-international-communityto-act-on-climate-change

IPCC (International Panel on Climate Change). (2014a). Summary for Policymakers. In O. Edenhofer, R. Pichs-Madruga, Y. Sokona, E. Farahani, S. Kadner, K. Seyboth, A. Adler, I. Baum, S. Brunner, P. Eickemeier, B. Kriemann, J. Savolainen, S. Schlmer, C. von Stechow, T. Zwickel, & J. C. Minx (Eds.), *Climate Change 2014, Mitigation of Climate Change. Contribution of Working Group III to the Fifth Assessment Report of the Intergovernmental Panel on Climate Change*. Cambridge: Cambridge University Press.

IPCC (International Panel on Climate Change). (2014b). Final draft Report, dated 17 December 2013, of the Working Group III contribution to the IPCC 5th Assessment Report "Climate Change 2014: Mitigation of Climate Change" Chapter 12, Human Settlements, Infrastructure and Spatial Planning. https://www.google.com.au/url?sa=t&rct=j&q=&esrc=s&source=web&cd=1&cad=rja&uact=8&ved=0ahUKEwjM4LyJ5rbSAhVBipQKHZjtBXIQFggZMAA&url=https%3A%2F%2Fwww.ipcc.ch%2Fpdf%2Fassessment-report%2Far5%2Fwg3%2Fipcc_wg3_ar5_chapter12.pdf&usg=AFQjCNHzV1daTw2fZRyO6hVamGghV0Q2uw&bvm=bv.148441817,d.dGo

Jones, S. (2012). A Tale of Two Cities: Climate Change Policies in Vancouver and Melbourne: Barometers of Cooperative Federalism? *International Journal of Urban and Regional Research, 36*(6), 1242–1267.

Jones, S. (2015). 'If You Can Make It Here, You Can Make It Anywhere': Performance Management and PlaNYC Climate Change Initiatives. *Regional Studies*. DOI 10.1080/00343404.2015.1052389.

Kennedy, C., Steinberger, J., Gasson, B., Hansen, Y., Hillman, T., Havráek, M., Pataki, D., Phdungsilp, A., Ramaswami, A., & Mendez, G. V. (2010). Methodology for Inventorying Greenhouse Gas Emissions from Global Cities. *Energy Policy, 38*(9), 4828–4837.

Kenny, M., & Cox, L. (2015). 'Bigger Threat than Terrorism': Barack Obama Signals Australia. India and China Must Improve on Climate Change. https://www.washingtonpost.com/politics/full-text-of-president-obamas-2014-address-to-the-united-nations-general-assembly/2014/09/24/88889e46-43f4-11e4-b437-1a7368204804_story.html?utm_term=.8ab354ab1953

Kimmelman, M. (2017). The Dutch have Solutions to Rising Seas. The World is Watching. *New York Times*, June. https://www.nytimes.com/interactive/2017/06/15/world/europe/climate-change-rotterdam.html?hp&action=click&pgtype=Homepage&clickSource=story-heading&module=second-column-region&region=top-news&WT.nav=top-news

Kolbert, E. (2015). The Siege of Miami. *The New Yorker*. December 21. http://www.newyorker.com/magazine/2015/12/21/the-siege-of-miami

Krause, R. M. (2011). Policy Innovation, Intergovernmental Relations, and the Adoption of Climate Protection Initiatives by US Cities. *Journal of Urban Affairs, 33*(1), 45–60.

Krause, R. M., Feiock, R. C., & Hawkins, C. V. (2016). The Administrative Organization of Sustainability Within Local Government. *Journal of Public Administration Research and Theory*. DOI 10.1093/jopart/muu032.

Lipsky, M. (1980). *Street Level Bureaucrats*. Nova York: Russel Sage.

Lipsky, M. (2010). *Street-Level Bureaucracy, 30th Ann. Ed.: Dilemmas of the Individual in Public Service*. New York: Russell Sage Foundation.

McCann, E. (2013). Policy Boosterism, Policy Mobilities, and the Extrospective City. *Urban Geography, 34*(1), 5–29. DOI 10.1080/02723638.2013.778627.

Mi, Z., Zhang, Y., Guan, D., Shan, Y., Liu, Z., Cong, R., Yuan, X. C., & Wei, Y. M. (2016). Consumption-Based Emission Accounting for Chinese Cities. *Applied Energy, 184*, 1073–1081.

Minx, J., Baiocchi, G., Wiedmann, T., Barrett, J., Creutzig, F., Feng, K., Förster, M., Pichler, P. P., Weisz, H., & Hubacek, K. (2013). Carbon Footprints of Cities and Other Human Settlements in the UK. *Environmental Research Letters, 8*(3), 035039.

Moynihan, D. P. (2009). Through a Glass, Darkly: Understanding the Effects of Performance Regimes. *Public Performance & Management Review, 32*(4), 592–603.

Nam, T. (2014). Assessing Operational and Collaborative Performance Management: A Case Study of PhillyStat. *International Journal of Public Administration, 37*(8), 514–527.

Obama, B. (2014). FACT SHEET: 16 U.S. Communities Recognized as Climate Action Champions for Leadership on Climate Change. https://obamawhitehouse.archives.gov/the-press-office/2014/12/03/fact-sheet-16-us-communities-recognized-climate-action-champions-leaders

OECD (Organisation for Economic Cooperation and Development). (2008). *State of the Public Service*. Paris: OECD.

OECD (Organisation for Economic Cooperation and Development). (2009). *Measuring Government Activity*. Paris: OECD.

OECD (Organisation for Economic Cooperation and Development). (2010). *Cities and Climate Change*. Paris: OECD Publishing. DOI 10.1787/9789264091375-en.

Osofsky, H. M., & Koven Levit, J. (2007). The Scale of Networks: Local Climate Change Coalitions. *Chicago Journal of International Law, 8*(2), 409.

O'Toole, L. J., Jr. (2000). Research on Policy Implementation: Assessment and Prospects. *Journal of Public Administration Research and Theory, 10*(2), 263–288.

Peck, J., Theodore, N., & Brenner, N. (2009). Neoliberal Urbanism: Models, Moments, Mutations. *SAIS Review of International Affairs, 29*(1), 49–66.

PLANYC (City of New York). (2007). PlaNYC: A Greener, Greater New York. New York City Mayor's Office of Long Term Planning and Sustainability, New York, NY.

Pope Francis. (2015). Final Declaration on Workshop on Climate Change. http://en.radiovaticana.va/news/2015/04/28/final_declaration_on_workshop_on_climate_change/1140356

Pressman, J. L., & Wildavsky, A. B. (1973). *How Great Expectations in Washington are Dashed in Oakland: Or, why It's Amazing that Federal Programs Work at All, this Being a Saga of the Economic Development Administration as Told by Two Sympathetic Observers who Seek to Build Morals on a Foundation of Ruined Hopes*. University of California Press.

Román, M. (2010). Governing from the Middle: The C40 Cities Leadership Group, Corporate Governance. *The International Journal of Business in Society, 10*(1), 73–84.

Rykkja, L. H., Neby, S., & Hope, K. L. (2013). Implementation and Governance: Current and Future Research on Climate Change Policies. *Public Policy and Administration, 29*(2), 106–130.

Schreurs, M. A. (2008). From the Bottom Up Local and Subnational Climate Change Politics. *The Journal of Environment & Development, 17*(4), 343–355.

SEF (South East Florida). (2012). Southeast Florida Regional Climate Change Compact. https://energy.gov/epsa/climate-action-champions-southeast-florida-regional-climate-change-compact-fl

Seto, K. C., Dhakal, S., Bigio, A., Blanco, H., Delgado, G. C., Dewar, D., Huang, L., Inaba, A., Kansal, A., Lwasa, S., McMahon, J. E., Müller, D. B., Murakami, J., Nagendra, H., & Ramaswami, A. (2014). Human Settlements, Infrastructure and Spatial Planning. In O. Edenhofer, R. Pichs-Madruga, Y. Sokona, E. Farahani, S. Kadner, K. Seyboth, A. Adler, I. Baum, S. Brunner, P. Eickemeier, B. Kriemann, J. Savolainen, S. Schlömer, C. von Stechow, T. Zwickel, & J. C. Minx (Eds.), *Climate Change 2014: Mitigation of Climate Change. Contribution of Working Group III to the Fifth Assessment Report of the Intergovernmental Panel on Climate Change.* Cambridge: Cambridge University Press.

Shigeto, S., Yamagata, Y., Ii, R., Hidaka, M., & Horio, M. (2012). An Easily Traceable Scenario for 80% CO_2 Emission Reduction in Japan Through the Final Consumption-based CO_2 Emission Approach: A Case Study of Kyoto-City. *Applied Energy, 90*(1), 201–205.

Sterck, M., & Scheers, B. (2006). Trends in Performance Budgeting in Seven OECD Countries. *Public Performance & Management Review, 30*(1), 47–72.

Taylor, D., & Balloch, S. (2005). *The Politics of Evaluation.* Bristol: The Policy Press.

UN Habitat. (2015). *Guiding Principles for City Climate Action Planning*, United Nations Human Settlements Programme (UN-Habitat). http://e-lib.iclei.org/wp-content/uploads/2016/02/Guiding-Principles-for-City-Climate-Action-Planning.pdf

UN-Habitat. (n.d.). *Urban Low Emission Development Strategies.* https://unhabitat.org/urban-initiatives/initiatives-programmes/urban-low-emission-development-strategies/

WHO (World Health Organisation). (2014). Ambient (Outdoor) Air Quality and Health, Fact Sheet 313. http://www.who.int/mediacentre/factsheets/fs313/en/

World Bank. (2010). *Cities and Climate Change: An Urgent Agenda*, Urgent Development Series, Knowledge Papers. Washington.

WRI (World Resources Institute). (2015). Global Protocol for Community Scale Greenhouse Gas Emission Inventories: An Accounting and Reporting Standard for Cities. C40 Cities and ICLEI. http://www.ghgprotocol.org/city-accounting

Young, A. (2007). Forming Networks, Enabling Leaders, Financing Action: The Cities for Climate Protection campaign. In S. C. Moser & L. Dilling (Eds.), *Creating a Climate for Change: Communicating Climate Change and Facilitating Social Change* (pp. 383–398). New York: Cambridge University Press.

3

City Governments, Climate Change, and the Performance Agenda

'Man plans, God laughs' (Mann traoch, Gott Lauch)
—Yiddish proverb

Introduction

There is a long running debate on the overall contribution of cities to global warming. Some scholars have a negative view. Many of the characteristics of cities such as higher consumption patterns, unsustainable land use management, and poorly designed transport systems are blamed for generating a significant proportion of the world's GHG emissions and contributing disproportionately to climate change (Sánchez-Rodríguez et al. 2005). These scholars consider the interactions between human activities in urban areas and the environment as negatively impacting on global warming. Other scholars adopt a more positive view and argue that the contribution of cities is often overstated. Dodman (2009) argues that there has been a failure to recognise that the varied consequences of global environmental change are likely to affect different urban areas in a variety of different ways. Satterthwaite (2008) argues that many of the

S. Jones, *Cities Responding to Climate Change*,
DOI 10.1007/978-3-319-64810-1_3

processes implicit in urbanisation can have a positive overall effect on global environmental change, as urban residents can generate a substantially smaller volume of GHG emissions than residents elsewhere in the same country. The argument here is that well-designed urban developments can reduce the impact of high population densities on global warming. Well-planned and well-governed cities provide the only hope of 'de-linking high quality of life from high levels of consumption', which is the critical factor with regard to reducing human contribution to global warming (Satterthwaite and Dodman 2009). The New Urbanism movement has established the theories and practices that have become a popular way to transform cities into 'green' sustainable living environments (Leccese and McCormick 2000). These arguments assume that the policies and planning practices of city governments in areas such as GHG mitigation have the potential to play a major role in reducing and/or increasing the impact of cities on global warming.

Context

Since the completion of the Kyoto Protocol negotiations in 1997, there has been a growing recognition of the role of city governments as critical players in the governance of mitigation and adaptation to deal with climate change (Kousky and Schneider 2003; Bulkeley and Betsill 2003; Betsill and Bulkeley 2007). Since that time, there has been a substantial amount of research examining the nature of the policy action taken by city governments. What emerges from the evidence is a mixed picture highlighting varying degrees of seriousness in the approaches taken. Some cities, such as New York, Stockholm, Copenhagen, and London, have developed and implemented substantial plans that reflect an important response to the climate issues they are facing. Others are labelled 'free riders', as they present little more than wish lists with statements of intent and encouragement that mirror the efforts of other cities but without commitments of altering practices or measurable implementation frameworks (Kousky and Schneider 2003). The approach taken by the Brisbane City Council (BCC) provides an interesting insight into how city governments can have the 'appearance' of taking action on climate change by having a plan, while actually doing very little (see Box 3.1).

Box 3.1 Brisbane City Council's Clean, Green WaterSmart City

The Brisbane City Council (BCC) is Australia's largest city government, as it covers the entire city and metropolitan area of Brisbane (1400 km^2). It serves a population of 1.2 million residents and has an annual budget of approximately $2.3 billion. In 2007, the BCC released its *Plan for Action on Climate Change and Energy*. The Council also released a *Clean Air Strategy* in 2009 with some mitigation elements included. By 2012, the BCC opted to incorporate all its plans, policies, and strategies under the broad umbrella of future priorities in an aspirational statement 'Brisbane Vision 2031'. The new programme titled 'Clean Green WaterSmart City' includes an amalgamation of strategies, plans, and policies for sustainability. The focus of the programme includes tree planting, purchase of carbon offsets and green power electricity, environmental rehabilitation, and stormwater drainage. When reporting on progress, the BCC provides public information through its Corporate Plan and Annual Report. The 2015 Corporate Plan outlines the BCC's objectives under each programme, with details of corresponding 'key actions'. For example, under the objective to reduce the Council's GHG emissions, the following details are provided:

1.1 Sustainability leadership

We will provide leadership and guidance to progress Council's sustainability agenda. We will maximise sustainability outcomes, reduce greenhouse gas emissions and support progress towards a carbon-neutral city.

Our medium-term objectives	Our key actions
Reduce Council's greenhouse gas emissions.	• Purchase 100% of Council's electricity from GreenPower sources. • Purchase accredited carbon offsets annually for Council's fossil fuel use. • Develop robust data capture procedures and an inventory of Council's carbon footprint.

When reporting on its performance against this objective, the BCC provides these details in its Annual:

Key results for 2013–2014

- 'Purchased 117,000 tonnes of accredited carbon offsets to neutralise diesel, petrol, and gas carbon emissions from Council's bus, ferry, and vehicular fleets and stationary gas use.
- Reduced Council's GHG emissions by approximately 86,100 tonnes a year by purchasing 100% GreenPower.
- Continued to retrofit Brisbane's streetlights, with 14,100 more energy-efficient lamps. Lamps will reduce electricity use by 40% over the next 2 years'. (BCC 2015)

The approach taken by the BCC allows business as usual in its operations. The Council essentially buys its carbon reductions, rather than improving efficiency or installing renewable sources of power, such as solar panels, to reduce demand from the grid. Source: Adapted from BCC (2015)

In most countries, city governments occupy a unique position in terms of climate change policy. Cities influence and control land use through zoning, official planning documents, building permits and development approvals, the supply and cost of parking, roads and public transit, and parks and recreational reserves. Some cities even possess regulatory and management roles in energy supply utilities. Research from 11 countries within the European Union (EU) reveals that if planned actions within cities were nationally representative, the countries investigated would achieve a 37% reduction in GHG emissions by 2050—a 27% reduction in GHG emissions for the entire EU (Reckien et al. 2014). The powers of Chinese city governments cover a wide range of emissions in the transport and building sectors, which together account for more than 60% of the nation's total energy consumption (Mai and Francesch-Huidobro 2014). Canadian research shows that, despite institutional limitations, city governments have significant potential through their existing powers and resources to make a major impact in mitigating GHG emissions. For example, Canadian local governments, including cities, have direct and indirect control over 52% of domestic GHG emissions (Robinson and Gore 2005). In California, cities with climate plans had more success in implementing strategies to reduce GHG emissions than cities without such plans. For example, they have more green buildings, spend more on pedestrian and bicycle infrastructure, and have implemented more programmes to divert waste from methane-generating landfills (Wheeler 2008).

Despite the great potential provided by their unique position, other reviews of city governments' policy suggests that while they claim to be taking more strategic and comprehensive approaches to climate change than other levels of government, in reality, many have been establishing strategies that are well beyond their legislated capacity (Bulkeley and Betsill 2005; Doucet 2007). Bulkeley and Betsill (2013) identify two categories of approaches taken by municipalities. The first group is 'characterised by individuals within municipal authorities recognising the potential significance of climate change and offering some form of response' (Bulkeley and Betsill 2013, 139). This approach, termed *municipal volunteerism*, involves pioneering cities with the resources and political will to take action on climate issues they regard as needing policy action. Cities in this category take an evidence-based approach to climate planning and policy, and are coincident with the broader direction of local governance within which 'accounting for performance was ever more important' (ibid.). In most cases, the political will was not strong enough to pursue regulatory approaches and thus most municipalities pursued enabling approaches that encouraged businesses and local communities to take action. In many cases, there was a gap between the rhetoric of municipalities and the level of response to governing climate change on the ground (ibid., 140). In other cases, city governments have been establishing targets aimed at reducing the overall GHG emissions within their jurisdiction that are more ambitious than those of their national governments. The second approach, labelled *strategic urbanism*, has seen a more political emphasis to its objectives. According to Bulkeley and Betsill (2013), this approach to urban climate change response involves new modes of governance. These modes tend to rely on different forms of 'partnership', or the 'blending of public and private authority, and a renewed interest in the ways in which both public and private actors might provide new forms of low-carbon and resilient infrastructure in cities' (ibid., 141). A central feature of this strategic approach is the use of alliances, both nationally and internationally, to establish networks, such as the ICLEI, C40, and the European Covenant of Mayors, that can apply political pressure on other levels of government to join municipal efforts to take action on climate change issues. The Paris Climate Agreement of 2016 contained new provisions outlining new voluntary reporting provisions for city governments and other non-state actors. The Non-State Actor Zone for Climate Action (NAZCA) is a Web-based portal that outlines the actions of cities and documents their commitments to reduce emissions. According to the NAZCA database, more than 400 cities and nearly 80 regions made formal emissions pledges. Unfortunately, there are a number of challenges ahead for the NAZCA before it can stand as an accurate and reputable vehicle for reporting. A recent survey of the material contained on the database revealed incomplete and incompatible data collection methods (Hsu et al. 2016). Unfortunately, this deficiency restricts attempts to measure and compare mitigation efforts. Without clear requirements, participants report data unevenly or idiosyncratically, often leaving out information on the sources and sectors targeted and the scope of particular initiatives.

However, the NAZCA is seen as an important step, and when greater efforts are made to track the outcomes of city based initiatives, there is some hope of identifying common benchmarks for tracking performance and implementation (ibid.).

The reasons why city governments pursue climate policy action vary widely, but there appear to be common themes behind policy choices. Krause (2011) has identified economic, political, quality-of-life, and psychological motivations for US cities adopting climate policy initiatives—see Table 3.1.

It is clear that many of the policy initiatives are voluntary and tend to be based on a recognition of urban and metropolitan areas' contribution to GHG emissions and an acknowledgement of the need for adaptation measures to protect city environments from potential impacts of global warming (Kennedy et al. 2012; Garren and Brinkmann 2012; McCarney 2013). Research examining the climate action by Quebec City (Canada) and Genoa (Italy) broadens the analysis and reveals a series of reasons why the two cities have been pursuing climate policy and how they intend to implement their strategies (Scanu and Cloutier 2015). Table 3.2 outlines the essential elements.

Climate change action plans established by city governments tend to reflect the institutional reality in which they are based. In multilevel systems, cities actions are constrained by their limited range of powers and the limited access to resources. Subnational governments largely shape city government roles and responsibilities. Research shows cities in the United States, Canada, and Australia are constrained by factors

Table 3.1 Motivations of local climate policy

Economic	Political	Quality of life	Psychological
Cost savings	External pressure	Improved air quality	Perceived risks/ vulnerabilities
Ability to attract external funding	Internal stakeholder pressure	Traffic reduction	Leading by example
Promote green jobs/local green industry	Credit claiming by elected officials in symbolic statements	Civic engagement and community building	

Source: Krause 2011 (adapted from Sippel and Jenssen 2009)

Table 3.2 Climate governance in Quebec City and Genoa

Factors shaping involvement
Ecological: geomorphological and climatic conditions
Economic: funds and subventions, growth opportunities
Institutional: normative frameworks and climate knowledge
Political: political will
Social: issues related to climate change
Framing: climate change as a policy problem
Dimensions of climate governance
Focus: adaptation and mitigation
Planning instruments: adaptation plan, mobility plan, greenhouse gas reduction plan, urban master plan
Sectors: energy, transport, water, waste
Modes of governance: self-governance, enabling, provisional, and authority
Actions: renewal of municipal fleet, rezoning, recycling, regulations, education campaigns
Actors: government, civil society, private sector, epistemic groups

Source: Adapted from Scanu and Cloutier (2015)

determined by state/provincial governments, including a limited range of legislative powers and restricted access to financial resources (Jones 2012). Similarly in the United Kingdom, the impact of the institutional context can be seen in the mitigation plans of city governments counteracted by factors beyond their control. As an example, the failure of the UK government to commit to complimentary legislation to support the emissions actions taken by the City of London has reduced the impact of the local measures—see Box 3.2.

European and US research provides clear evidence that political leaders also face important challenges in their approaches to climate policy. Given the high degree of politicisation around government action on climate change, there is evidence suggesting a polarisation of opinion falling along politically partisan lines (Eurobarometer 2009; McCright and Dunlap 2011). Liberal and left-of-centre political parties with leaders prepared to champion a climate agenda increase the possibility of a policy window emerging (Dietz et al. 2007). According to Kingdon (2003), the view of the political leader alone may not be enough to drive policy change. In the case of city governments, other factors are required to propel a commitment to a policy response to climate change. Kingdon argues that a policy window emerges when a suitable policy response is

Box 3.2 Attempting to reduce emissions in London

Despite having a mitigation plan that includes targets to reduce greenhouse gas (GHG) emissions from traffic through initiatives such as a city congestion charge, London emissions continue to grow. To further improve the air quality, Mayor Boris Johnson also plans to start charging drivers of older diesel cars $19 extra to enter the city by 2020. Even with such measures, it is the impact of surrounding regions that largely determines the emissions levels in the city, and these are shaped by UK government's failure to take serious action on emissions from traffic in the national context. According to European Union (EU) data, London has the worst nitrogen dioxide level in the EU. As a result, the EU took a serious step to generate action by the national government. In 2015, the EU commenced proceedings against the United Kingdom for its 'failure to respond to 15 years of warnings to cut high levels of nitrogen dioxide air pollution, which is mostly caused by traffic' (Nelsen 2017). European Commission research estimates suggest that the EU-compliant level of air pollution in London will not be achieved until 2025.

available and a problem or crisis exists that warrants policy action by the political leaders (2003, 113). These three 'streams' will need to be 'coupled' to open an opportunity 'window' for a government to respond. An example of Kingdon's perspective is, first, a popular leader emerges with a mandate to take action on climate change. Second, there is evidence that raises public opinion that something needs to be done about reducing emissions in the city, health risks, for example. And third, the city government has the legislative authority to introduce policies that have been shown, through scientific evidence, to reduce emissions. For Kingdon, the three scenarios are a 'coupling' of three streams, 'problem', 'policy', and 'politics' establish a policy window for policy action to be taken. For Kingdon, a policy entrepreneur is needed to drive the policy within the government. In the case of city governments, the mayor or governor as the political leader is often the most obvious policy entrepreneur (Zerbinati and Souitaris 2005; Sharp Daley and Lynch 2011).

A survey of over 50 US mayors of large cities reveals that those with affinity to the Republican Party tend to be critical of government developing and implementing climate-related policy. Republican mayors prefer to focus on more traditional local government responsibilities such as economic growth and infrastructure development. The survey, conducted by Boston University, found most notable distinctions between Republican and Democratic mayors on their views on the role cities should play in combating climate change. Results show that mayors of different parties express sharply different views that mirror national politics. While less than one third of Republican mayors believe that cities should expend resources to mitigate the risks of climate change, approximately 9 in 10 Democrats supported greater investment in mitigation and adaptation initiatives (Einstein et al. 2014, 3). Australian survey research also shows that 'progressive' politicians are more likely to believe in climate change and pursue policy action (Fielding et al. 2012).

In some contexts, climate action by city governments can induce further action by other levels of government. For example, recent efforts by the city government in Delhi to reduce emissions by reducing vehicle use will be made more effective through complimentary measures by other levels of government (see Box 3.3). The policy action by the Delhi government has been a catalyst for other institutional stakeholders to take action such as the courts and other levels of government. As a result, there are predictions that a considerable reduction in emissions in the city with the world's worst air quality will be achieved by 2020. Delhi has

Box 3.3 A bold experiment in Delhi

In January 2016, the Delhi local government announced a two-week experiment to reduce car emissions by restricting road use to odd or even numbered licence plates on alternate days. Experience of other cities with similar measures, such as Beijing and Sao Paulo, raised doubts that the experiment would work. Results have shown, however, that over the two-week period, in Delhi, emissions were down from their usual extremely high levels. The success was achieved through the combination of support from volunteers shaming shirkers into parking their cars, police out in force to fine those driving on the wrong day, and more people choosing public transport. As a result of this city government success, the Supreme Court has banned all luxury diesel cars from the city, and the national government agreed to introduce stricter emissions standards for passenger cars and improve the quality of fuel. Other government regulations will contribute to reducing emissions, for example, expanding Delhi's metro, improved road maintenance, restrictions on heavy vehicles entering the city, and enforced retrofitting of older vehicles (*The Economist* 2016).

special constitutional status within the Indian political system, making it difficult to serve as a model for other Indian cities. Nevertheless, the action taken to ban diesel cars is one being considered by other major cities such as London, Mexico, Madrid, Athens, and Paris (Harvey 2016). Chinese cities face considerable constraints in their institutional context from central agencies, yet cities such as Hong Kong, Shenzen, and Guangzhou have distinctive features that allow them a high degree of vertical autonomy in policy-making, planning, and different forms of governance (Mai and Francesch-Huidobro 2014, 8).

Research scholars in the United States observe that, notwithstanding the restrictions imposed by the factors outlined above, many city governments have plans that set emissions reduction goals, establish emission inventories, enable green public sector operations, and recommend a range of other measures (Wheeler 2008). Many climate plans have been developed through extensive stakeholder processes and present very detailed lists of recommendations with quantified emissions benefits. The overarching challenges for all city governments are designing policies that reflect community capacity to make the changes necessary to mitigate climate change and introducing adaptation measures to manage the environmental impact brought about by global warming. Policies will require lifestyle changes by citizens and the introduction of sustainable development practices. It would appear from the policy work already done that many city governments are well advanced in these areas. One example is the Australian city of Adelaide with its *Energy Management Action Plan*, which the city government argues has contributed to a 19% reduction in GHG emissions in the city since 2011 (ACC 2011). The Adelaide example reveals the importance of a supportive business community, acceptance of change within the wider community, and complimentary state government legislation that has supported the development of renewable energy (see Box 3.4).

Despite the existence of some highly successful examples, concern persists within the policy community over such varied approaches by city governments to climate policy development and implementation. Some have well-developed climate change policy where urban climate governance arrangements set the climate action agenda and promote implementation schemes. Others have neither well-developed urban

Box 3.4 Adelaide Energy Management Action Plan 2011–2014

Adelaide is the fifth-largest city in Australia and the capital of South Australia (SA). In 2014, the Adelaide metropolitan area had an estimated population of 1.29 million. The Adelaide City Council (ACC) is responsible for the City of Adelaide, with a population of approximately 23,000. Since 1994, the ACC has reduced its own emissions by 60%. The *Energy Management Action Plan 2011–2014* 'focuses on embedding sustainability in the organisation to reduce energy use and carbon emissions and assisting the community to be an energy efficient city' (ACC 2011). The *Carbon Neutral Council Action Plan 2015–2025* (the Plan) was introduced in 2015 to support emissions reductions, with an aspiration to make the city carbon neutral by 2020. The ACC's initiatives are supported by the SA government's *Climate Change and Greenhouse Gas Emissions Reduction Act 2007*, which legislates targets to reduce GHG emissions in the state by 60% by 2050. There is a joint agreement between the state and the ACC to reduce emissions in the city. The Plan details specific objectives, including improving energy efficiency, increasing local generation of renewable energy, using accredited carbon offsets, diversifying fuel sources, organising community engagement programmes, and facilitating behaviour change. Mitigation measures are line items in the ACC's annual budget. There is also a series of key performance indicators relating to the ACC's own energy use and attempts to support and influence community energy management. There are details on actions taken over the 2011–2014 period, including the action's relevance to the objectives, the budget allocation, time frames, the level of priority and responsibility for implementation. One example is the description provided in the Plan for the measure 'To reduce energy use by 15% and carbon emissions by 30% from buildings', which appears as follows:

Description of Action	Reducing Emission	Using Renewable Energy	Offsetting Emissions	Budget (*$ '000s*)	Timeframe	Priority	Responsibility
Strategy 1: Reduce energy use by 15% and carbon emissions by 30% from buildings (cont'd)							
1.2 **Develop sustainability minimum standards and guidelines for all major strategic projects and assets managed by Council.** The guidelines would integrate energy use, emission reduction principles and financial criteria into the planning, design and construction process of all new Council assets and retrofit/refurbishment projects.	x	x		(15*)	Year 1 and ongoing	High	**Strategic Asset and Property** **Sustainable City** Capital Works and Infrastructure Management

'The ACC provides public reports on the progress of each its strategic outcomes through annual reports. In 2014 the ACC and the State Government in a joint announcement claimed the city's community had reduced emissions by 19%. This was largely the result of complimentary measures and the actions of a coalition of local stakeholders. The ACC's Programme Manager for City Sustainability claimed, 'Results like this demonstrate the benefits of long term planning, industry leadership and a commitment to implementation, South Australia and the City community, particularly commercial property owners, have demonstrated that renewable energy and green buildings are highly effective strategies to reduce carbon emissions'. (ACC 2014) governance arrangements nor comprehensive climate policies (Lee and Painter 2015). Some city governments restrict their options to govern through 'enabling' by providing information and guidance to the community on energy efficiency and emissions reductions. This is a very common approach, as it allows governments to argue that they are doing something about climate change by setting benchmarks and consistent approaches without needing to commit substantial resources or unpopular regulations (Kern and Abler 2009). When city governments govern through laws and regulations, they are displaying a more serious commitment to action than through enabling and provisional methods. By passing laws and regulations that support mandatory climate change initiatives in areas such as urban planning, energy supply, and transport infrastructure, city governments send a strong message about their commitment to action on climate change (Engel 2009). It is the inconsistency in approaches taken by cities in the same country and even the same state/province that raises concerns in the policy community. The inconsistency also filters through to measurement and evaluation, with many local governments more focused on demonstrable action, such as announcing a plan, building bike paths, or planting trees, than on applying cost–benefit analysis to proposals with corresponding indicators and milestones to evaluate progress towards GHG reduction targets.

A variety of factors, including the impact of interest group pressure, political institutions, and problem severity, can influence a city's decision to develop and implement climate protection programmes, to name a few. In the past, local mitigation policy has been predominantly a top-down decision based on what elected representatives believe to be good business or rational economic and political choices. Emissions reduction goals vary widely, many proposed actions are voluntary, and few resources have been allocated; implementation of most measures is slow or has not yet taken place. Researchers argue that most plans do not address adaptation to a changing climate. Evidence reveals that 35% of European cities studied have no dedicated mitigation plan and 72% have no adaptation plan (Reckien et al. 2014). Kousky and Schneider (2003) examined the climate policies of 23 US cities and found that in the majority of cases, policy has not been driven primarily by widespread public pressure, nor wholly for climate protection, but instead, justified by cost savings and other perceived co-benefits. This is especially true for co-benefits, as local governments do not typically attempt to value or quantify them. Other research on Indian cities by Revi (2008) reveals that there will need to be greater focus on adaptation to cope with the anticipated impacts of extreme weather events that continue to impact on different areas. The primary challenges in the Indian context are better and more accountable city governance, democratic decentralisation, and improvement in the functioning of public institutions. There is cause for optimism in many countries, where city officials are witnessing a rapidly growing public awareness of the climate issue and general support for climate change planning; however, there is also reluctance to change personal behaviour (Wheeler 2008).

Trends and Issues

What trends and issues does current research observe in establishing climate policy, implementation, and measurement at the city level? First, there is some evidence that city governments can be effective in mitigating emissions through the implementation of their climate action plans. Data from Berlin, Boston, Greater Toronto, London, New York City, and Seattle for the period 2004–2009 revealed that all were reducing their per capita GHG emissions in accordance with their respective climate action plans (Kennedy et al. 2012). The actions taken by these cities have been shown to be reducing emissions at a faster rate in percentage terms than the subnational and subnational governments within their respective jurisdictions (Kennedy et al. 2012). Similarly, Hong Kong, Shenzhen, and Guangzhou city governments have been proactive in the sectoral integration of carbon reduction and low-carbon innovations (Mai and Francesch-Huidobro 2014, 8). The experience of these city governments suggests that the development and implementation of climate change plans with well-targeted, specific, achievable, and measurable actions are important contributors to effectively reducing emissions.

Table 3.3 GHG-reducing activities of US cities

Eleven most popular GHG-reducing activities	Percentage of cities adopting (n = 329)
Curb-side recycling provided for residents	91.18
Greening of the vehicle fleet	80.85
Information for residents on energy efficiency	76.83
Bicycle pathways	69.51
Planning and zoning decisions consider effects on sprawl	66.26
Community-wide hike and bike trails	64.63
Residential green waste composting	62.8
Efficient lighting in city buildings	59
Public transportation	90.98
Education and outreach regarding privately owned trees	56.10
Climate protection coordination unit established	55.79

Source: Adapted from Krause (2011)

Second, city governments with insufficient political will or capability to prioritise their climate actions tend to become ineffectual in achieving their targets (Erickson et al. 2013). These cities often establish plans because they believe that it is reasonable and appropriate for a city government to have a plan to do something about climate change. City leaders tend to focus on the economic co-benefits and on initiatives that reduce government operating costs (e.g., energy efficiency measures) or even generate revenue (e.g., traffic tolls). Krause (2011) surveyed over 300 US cities looking at the primary actions taken in their climate change plans. Table 3.3 summarises her results, which reveal the dominance of economic factors and co-benefits of climate action being the major focus. Close examination of the most popular measures reveals a list of typical city government responsibilities not necessarily unique to climate change action. Krause (2011) regards these as 'easy policies' or the 'low-hanging fruits' that councils will naturally attempt first before moving on to more challenging policy measures. Other research shows that climate measures are often unlikely to lead to fundamental changes in the operations of city governments. Millard-Ball claims that the city government staff he interviewed 'felt that they rarely needed to invoke a climate plan in order to advance a particular policy, as their colleagues and elected officials wanted to implement "green" measures regardless of the contents of a plan' (2013, 302).

The second possible reason relates to access to data and the relevant accounting tools. Research suggests that urban planners and local decision-makers often lack the tools and means needed to make informed choices about the climate change implications of local growth and redevelopment decisions or to measure the effects of their decisions. While a substantial number of GHG accounting tools currently exists, few have the capacity to work at both the regional and the local scale or to capture the multiple consequences of regulatory decisions. Researchers argue that most local governments do not have access to GHG data in a way they can use it to translate into the areas of responsibility, such as land use planning (Miller et al. 2009). Without this data, it is challenging for city governments to demonstrate the effectiveness of their climate policies.

A third observation is that many city governments argue that they have unique circumstances, including development histories, economic structure, urban form, and institutional and financial capacities among other factors (Solecki et al. 2015), that form path dependency to shape the specific blend of policy initiatives that make up their climate plans. As an example, research examining the factors determining the development of GHG inventories in seven global cities—Bangkok, Chicago, London, Madrid, Mexico City, Milan, and New York City—shows that decisions ultimately depend on economic, technical, social, and cultural factors (Croci et al. 2011). Factors considered unique to each context play a role in shaping the depth and extent of inventories. These can include urban density, electricity consumption patterns, and technological features of power generation. However, urban forms affecting mobility patterns and technological features of the vehicle stock stand out as the most significant determinants (ibid.). As a result, one quarter of European cities surveyed have both an adaptation and a mitigation plan, and set quantitative GHG reduction targets, but those vary extensively in scope and ambition (Reckien et al. 2014). US evidence shows that organised interests also influence both the adoption and the implementation of climate mitigation programmes. This effect, however, is contingent on political institutions. In the United States, organised interests in the local context are more effective in mayoral as opposed to city manager forms of governments. Interestingly, while financially constrained cities may adopt climate mitigation programmes to advance co-benefits or cost savings, fiscal stress can also impede programme implementation (Sharp et al. 2011).

Fourth, in remarkably few cases, city governments will establish a climate action plan with an implementation process that includes a review and evaluation framework. Good policy practice shows that implementation includes establishing review and evaluation processes that help identify areas for improvement and new policy opportunities (Sanderson 2002: McConnell 2010). It seems that few cities use the results of the initiatives taken, in terms of GHG emissions targets and improving the city's resilience, to pursue policy and management change. For example, research by Sippel (2011) reveals that most German cities do not use their GHG reporting and emission targets as meaningful GHG management tools. The majority of targets are not city specific, and almost half of the cities do not report base-year emissions. As a consequence, no single Western German city could demonstrate that it was on course to reach its emission target (Sippel 2011). Research on cities in Florida reveals a range of issues in the use of GHG inventories as a key policy and management tool. Local government staff involved in developing measurement methodologies of emissions in the different cities applied different approaches. A review of methodologies revealed a need to link GHG emissions to policy, and to improve methodologies and the data collection process. More consistent approaches would provide policy learning opportunities and improvements for tracking the effectiveness of specific GHG policies (Garren and Brinkmann 2012). A study of US cities in the ICLEI network revealed that for about 50%, there were no updating standards nor provisions or mechanisms to ensure updating of policy initiatives based on the experiences of other cities. Most importantly, from a network coalition perspective, there were no provisions for policy learning. Information sharing between practitioners regarding lessons learned in tracking data from utilities, various government departments, and other entities is ad hoc (Blackhurst et al. 2011).

Finally, researchers note a lack of rigour by city governments in measuring emissions. Individual cities tend to more frequently take estimates largely based on energy use and CO_2, not GHG, emissions. Researchers argue that a fuller examination of emissions from additional GHGs and sectors would improve our understanding of cities' contribution to global climate change (Marcotullio et al. 2013). Most case studies focus on similar types of urban areas such as large cities (Kennedy et al. 2012; Butler et al. 2008). This leaves few estimates of emissions from medium-sized and small centres, where urbanisation currently occurs most rapidly. In a working group report, the IPCC (2014) revealed a number of knowledge gaps that have been preventing the development of measurement systems that accurately reflect the impact of mitigation schemes by city governments. Of particular concern is the lack of consistent and comparable emissions data at the local level, making it particularly challenging to assess the urban share of global GHG emissions as well as to develop urbanisation typologies and their emissions pathways. In addition, there are few evaluations of urban climate action plans and their effectiveness. Despite the presence of plans and strategies, researchers argue that the actions by European cities would often be insufficient to reach national targets and fall short of the 80% reduction in GHG emissions recommended to avoid global mean temperature rising by 2°C above pre-industrial levels (Reckien et al. 2014). Importantly, there is a lack of consistency and thus comparability in local emissions accounting methods, making cross-city comparisons of emissions or climate action plans difficult (Seto et al. 2014, 929). A study of 40 German cities found that all had GHG inventories or were preparing inventories, but unfortunately, comparability of reporting data between cities is limited due to varying methodologies and the frequency of reporting (Sippel 2011). Such measurement issues are worrisome, as they suggest either a serious lack of concern or an inability to demonstrate success.

These observations raise questions about the importance of city governments' ability to measurably demonstrate the success of their climate action plans. This is particularly true in terms of the GHG emissions reduction targets, as these tend to be the focus of national and international efforts. Most urban responses to climate change worldwide have focused on the mitigation of GHG emissions. The UNFCCC came into force in 1994, with the aim of stabalising GHG concentrations in the atmosphere by reducing the impact of human activity on the climate (UN 1992). Deploying a monitoring system that offers sustained, accurate, transparent, and relevant information represents a challenge for GHG accounting and public policy (Duren and Miller 2011). The importance of measurable demonstrations of effectiveness fits into the agenda of city leaders, and networks and coalitions of city governments to raise the profile of cities and the legitimacy of their claims to be serious stakeholders in climate negotiations.

There are many typologies and frameworks that could establish comprehensive approaches to the development and implementation of climate action plans by city governments. At present, research focuses on developing improvements to establish inventories to support improved measurement of GHG emissions and help satisfy the concerns of the IPCC. These are important elements of a performance measurement scheme, as they provide the critical measurement components that serve as a baseline for more effective performance management systems. However, the performance management literature stresses the importance of embedding performance in the management processes and procedures of city governments (Sanger 2008; Ammons 2013). The presence of high-quality data alone is insufficient in itself to improve policy implementation and the delivery of climate policy outcomes. The collection of data alone does not automatically lead to improvements in policy implementation. Surveys of practitioners and researchers identify that it is essential to 'understand what happens to performance information after it is reported and evaluated' (Kroll 2013). Public managers argue that performance information alone will not help them without the resources to do what needs to be done (Hatry 2006).

The public policy literature stresses a focus on the uses to which performance data is applied. This can be the critical element in the implementation of policy (Marr 2009). The use of performance data is critical in terms of reporting and management change. Improvements in the measurement and management of performance hold some promise for city governments serious about measurably demonstrating the success of their climate change plans. A central tenet of the performance literature is that an effective performance management framework will provide public officials with the necessary tools to manage their programmes and allocate resources that will lead to improvements in policy implementation and the delivery of policy outcomes (Moynihan et al. 2011). If we assume, based on OECD (2009) research, that performance management enables governments to quantify promises, and measure and report their actions, then city governments need to enable citizens, managers, and politicians to make meaningful decisions about increasingly complex climate change policy implementation issues. Sanger's (2008) assessment of case studies involving cities that have sought to develop and use performance measurement and to engage citizens demonstrate great potential to improve governance and manage for results that citizens want. Ammons (2013) also found evidence, in a review of budgets and other performance documents of cities at the forefront of performance measurement, of advances in the reporting of service quality, efficiency, and effectiveness measures. The City of Stockholm provides an example of the way performance information can form part of an integrated approach to the implementation of climate initiatives by a city government. In both its national and international contexts, Stockholm has been using data as both a management tool and as evidence for recommendations for policy change (see Box 3.5).

A Role for Performance Management

Improvements to performance measurement and reporting by city governments hold considerable importance in the policy implementation agenda for achieving outcomes. There are considerable implications for the climate change agenda being set by city governments nationally in their multilevel systems and internationally as established networks and coalitions to support their actions. In the context of the claims being made by city governments and from the ambitious targets and proposed actions, it is critical that they demonstrate their success through the application of clear and objective measurements. Observations in the literature indicate that there are three key interrelated agendas driving the need for improvements in city governments' capacity to demonstrate their success in implementing climate policy.

The approach taken by Stockholm represents a response to what are, arguably, three fundamental performance agendas behind the need for city governments to measurably demonstrate their success with climate policy. By working in conjunction with the Covenant of Mayors, Stockholm is addressing the *international agenda*, calling for cities to

Box 3.5 Stockholm: A fossil fuel–free city by 2050

The city government of Stockholm established its first climate plan in 1998. In 2009, the Mayor signed the European Mayors' agreement 'Covenant of Mayors' to work with other cities to reduce emissions by 20% by 2020. As part of the Covenant, members need to submit a report every two years regarding the implementation of their plans; the report enables evaluation and review of actions taken. Stockholm argues that it is exceeding the requirements of the Covenant, having reduced its emissions by 23% in 2015. In terms of performance management, Stockholm's approach is that the city connects its environment programme, city plan, and energy plan to constrain the actions that can be taken by committees, city boards, and the city council. The *Energy Action Plan* is a guiding document in terms of management and budget decisions and the operational planning across the city government. The measurement activities fall across each major initiative in the Plan. Public reporting provides details of each initiative, such as the objective to expand the number of green vehicles in the city as detailed below:

Measure	Timetable	Year	Reduction of greenhouse gases (CO_2e)	Energy efficiency
Increase the number of green vehicles	• 50 per cent of newly registered private cars should be environmentally classified vehicles by 2014.	2015	54,600 tonnes	195.6 GWh
		2020	153,000 tonnes	548.3 GWh
	• 100 per cent of newly registered private cars should be powered by alternative fuels by 2020.	2030	535,500 tonnes	1,673.2 GWh

Assumptions for the calculations

- 50 per cent of new car registrations in 2014 are green cars, small change in the type of cars being bought.
- 100 per cent of new car registrations in 2020 are green cars.
- 12.5 per cent of the vehicle park is renewed every year, of which half of the green cars remain in Stockholm. This means that there is an relocation of green vehicles from the city at a rate of 50 per cent of the rejuvenation.
- Increased low-concentration inmixing to 6.5 per cent ethanol
- When it comes to the distribution of vehicles, consideration has been taken to the EU's suggestion regarding demands that emissions from newly registered cars can be no higher than 95 grams of CO_2 per kilometre from 2020 and 70 grams of CO_2 per kilometre in 2025. (The Swedish Transport Administration, Traffic-type-independent planning basis for limited climate impact, Publication 2010:095)

Finance	The cost of participating in and running demonstration projects depends on the financing of each separate project. In certain cases there is an opportunity for seeking support.
Responsible	The City of Stockholm is responsible for promotion and the actors for execution.

Source: City of Stockholm (2012, 33)

The public can see the detailed assumptions made about the measures used to determine progress and success of the initiative. Time frames and anticipated emissions reductions as a result of the initiative are also provided. The city government regards these as knowledge-enhancing measures that will inform future decisions on new proposals on climate and energy goals. As an example, the measurement report analyses car sales and fuel types. By 2015, results were leading the city government to make recommendations to the national government that its definitions of fossil-free needed to focus more on energy-efficiency, and to increase the extent of vehicles powered by renewable fuels; otherwise, the 'fossil-free' targets would not be reached.

implement uniform approaches to measuring and reporting to help legitimise the role of city governments in international and national negotiations on climate change action. Second, Stockholm, through its measurement scheme, is in a position to justify its action and provide evidence of effectiveness, which helps meet the *community engagement agenda*. With performance-based evidence, Stockholm is in a better position to engage with citizens regarding behaviour change required for climate action. The third agenda, *organisational governance agenda*, of embedding sustainability and climate change considerations into the governance arrangements in city governments' operations, is addressed by Stockholm's use of performance measurement as evidence for behaviour change within its own management and policy systems. In order to understand more comprehensively the significance of the three agendas, the next sections will address each in turn.

International Agenda

With the international agenda, the global challenge is to establish a set of standardised international measures and indicators that monitor the impacts of climate change on cities. For some, the universal indicators form part of the objectives of organisations such as the ICLEI, Partners for Climate Protection, and C40 Cities projects to help legitimise city governments as actors in international negotiations on action over climate change. McCarney (2012) advocates the development of indicators that can empower decision-making and build capacity within city governments so they can perform as new sites of governance in global negotiations on climate change. In 2014, the McCarney's *Global City Indicators Facility* in Canada was successful in establishing ISO 37120 as a certification programme that collects a wide range of data on the quality of life and environmental health in cities. The standard is designed to produce consistent urban-based data that is verifiable by third parties. Such initiatives help to improve the reputability of and the confidence in data that can be used for measurement and monitoring as well as to enhance policy-making (Kitchin et al. 2015).

A study of over 250 members of the ICLEI city network examined the members' achievement of the five-stage process of the implementation of climate plans, including: (1) conduct a local GHG emissions inventory, (2) adopt emissions reduction targets, (3) develop a local climate action plan, (4) implement the plan, and (5) monitor and verify the results. Of all the cities surveyed, 110 (42%) had joined the ICLEI but had not completed any milestones, 36 (14%) established a GHG inventory, 58 (23%) developed a climate action plan, 31 (12%) implemented their plans, and 21 (8%) met the final milestone of evaluating the local climate plan—see Fig. 3.1.

Only cities reaching Stage 5 could be considered to be making serious progress towards reducing GHG emissions (Reams et al. 2012). This

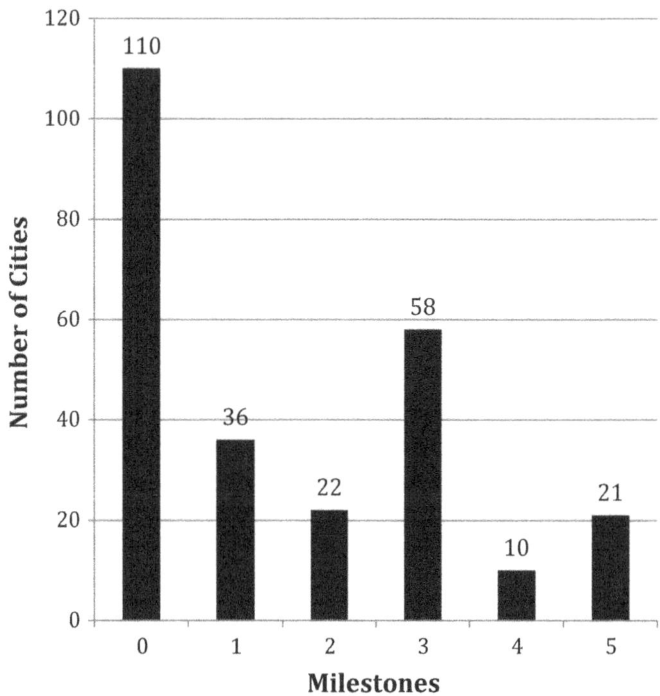

Fig. 3.1 Completion of the five-stage implementation of climate plans among 257 ICLEI members
Source: Adapted from Reams et al. (2012, 140)

quantitative research is important because it raises awareness and provides evidence of the fact that many city governments are satisfied by baseline volunteerism by simply registering as members of the ICLEI and similar networks but not pursuing policy action on climate change.

This reluctance to implement and evaluate has driven many researchers to look for possible reasons why these governments have been unable or unwilling to take more serious action on climate change. City government officials willing to discuss the main issues frequently cite political tensions as a prime reason why many unpopular measures, for example, those that would increase household costs, tend to languish (Wheeler 2008). Many officials view the creation of emissions inventories and climate change plans as significant achievements in themselves (ibid.). Moving beyond establishing the technical and administrative aspects of climate plans proves difficult for cities that are facing resource constraints, where more immediate policy issues take priority.

A key challenge for many city governments has been the establishment and maintenance of a comprehensive GHG inventory. Further assessments of ICLEI members shows that the implementation of a climate policy that achieves meaningful reductions will involve careful analysis, skilled decision-making, and dedicated effort (Wheeler 2008). There have been calls for a paradigm shift from standards based on a traditional and simple calculator method to more comprehensive approaches that facilitate decision-making and improve climate action plan implementation. Researchers have identified a need to improve the capacity of city governments to maintain their inventories and keep them updated. Information sharing for policy learning has also been a recommendation of researchers looking for ways to assist cities to overcome resource deficiencies. By learning from the experiences of others, city governments can avoid costly mistakes and adopt policies that have proven to be effective in other, similar, contexts (Blackhurst et al. 2011).

Both the ICLEI and the C40 are among the international network coalitions focused on the objective of developing measurement systems that monitor the impact of city governments. There have been mixed results in terms of the individual cities that have adopted and attempted to implement the systems developed by these organisations (Wheeler 2008; Roman 2010; D'Avignon et al. 2010; Brander et al. 2013). As we have seen above, there is evidence of some successful adoption of ICLEI's protocols and the influence of various factors that contributes to implementation. Research shows that these programmes require time to fully mature, perhaps reflecting the degree of organisational investment and change required (Wheeler 2008). In 2015, the C40 established a series of pilot studies with approximately 35 cities across the world. The cities are from both developed and developing nations, and with varying degrees of experience in climate policy. Each has committed to implement the *Global Protocol for Community-Scale Greenhouse Gas Emission Inventories* (GPC), established jointly by the ICLEI and the WRI, which focuses on '[p]roducing tools, standards and frameworks to support cities to implement the most impactful mitigation and adaptation actions and measure and manage their effectiveness' (C40 2016). The C40 also has the support of the World Bank, UN-Habitat, and the United Nations Environment Programme (UNEP) in promoting the protocol. The intention of the GPC is to move towards a standardisation of the measurement of emissions. The President of the WRI, Andrew Steer, claims:

> If we want to turn the tide against climate change, cities will need to lead the way. Compact and efficient cities can dramatically reduce emissions and will drive innovation and sustained economic growth. Until recently there has been no consistent way to measure city-level emissions. Now, that has changed. We now have a common international standard to inform strategies to cut emissions and create better, more livable cities. (C40 2014, 1)

The C40 claims that its inventory methodology allows for the aggregation of emissions at the regional and national levels. By following the sources and scopes detailed in the GPC, see Fig. 3.2, the C40 argues that city governments will improve the data quality for reporting purposes, measure the contribution of their mitigation actions to regional and national targets, and identify innovative transboundary and cross-sectoral strategies for mitigation. Inherent within this approach is the long-term view of the C40 that the GPC will contribute to the global-level challenge to establish a set of standardised international measures and indicators that monitor the impacts of climate change on cities. With a standardised approach, these networks argue that they establish the

Sectors and sub-sectors	Scope 1	Scope 2	Scope 3
STATIONARY ENERGY			
Residential buildings	✓	✓	✓
Commercial and institutional buildings and facilities	✓	✓	✓
Manufacturing industries and construction	✓	✓	✓
Energy industries	✓	✓	✓
Energy generation supplied to the grid	✓		
Agriculture, forestry, and fishing activities	✓	✓	✓
Non-specified sources	✓	✓	✓
Fugitive emissions from mining, processing, storage, and transportation of coal	✓		
Fugitive emissions from oil and natural gas systems	✓		
TRANSPORTATION			
On-road	✓	✓	✓
Railways	✓	✓	✓
Waterborne navigation	✓	✓	✓
Aviation	✓	✓	✓
Off-road	✓	✓	
WASTE			
Disposal of solid waste generated in the city	✓		✓
Disposal of solid waste generated outside the city	✓		
Biological treatment of waste generated in the city	✓		✓
Biological treatment of waste generated outside the city	✓		
Incineration and open burning of waste generated in the city	✓		✓
Incineration and open burning of waste generated outside the city	✓		
Wastewater generated in the city	✓		✓
Wastewater generated outside the city	✓		
INDUSTRIAL PROCESSES AND PRODUCT USE (IPPU)			
Industrial processes	✓		
Product use	✓		
AGRICULTURE, FORESTRY AND OTHER LAND USE (AFOLU)			
Livestock	✓		
Land	✓		
Aggregate sources and non-CO_2 emission sources on land	✓		
OTHER SCOPE 3			
Other Scope 3			

✓ Sources covered by the GPC

● + ● Sources required for BASIC+ reporting

Sources included in Other Scope 3

● Sources required for BASIC reporting

● Sources required for territorial total but not for BASIC/BASIC+ reporting (*italics*)

● Non-applicable emissions

Fig. 3.2 Sources and scopes covered by the GPC

legitimacy of city governments in national and international negotiations on climate initiatives. Legitimacy was raised as an issue of concern in Copenhagen by city mayors in 2009 when they held their own *Climate Summit for Mayors*.

The C40 provides examples where city governments have been implementing the GPC, including the following:

- Guangzhou, China, is using the GPC to analyse its GHG emissions trends and design a road map towards emission peak. WRI China provides training and technical advice to Guangzhou to apply the GPC.
- Rajkot, India, and seven other Indian cities set up the first GHG inventories using beta versions of the GPC. The GPC guidance has helped Rajkot (one of the GPC pilot cities) plan and implement actions to achieve its 14% CO_2 emissions reduction target by 2016 (based on the 2011 base year).
- Wellington City, New Zealand, participated in the GPC pilot programme to develop a GHG inventory for the Wellington Region (including Wellington City and seven other cities) as part of its new climate change action plan to lower GHG emissions by 30% by 2030 and 80% by 2050 (from 2001 levels).

The research literature illustrates an international trend of city governments engaging in some way with the challenges posed by climate change. The capacity of these governments varies considerably, and researchers and city networks are attempting to identify solutions to many of the constraints currently impacting on implementation. One key area of research activity is the development of universally applicable metrics that can monitor the impact of the actions being taken by cities across the world through their climate plans. At the heart of the measurement solutions being developed is the need for high-quality comparable data. However, there is some hesitation by many city governments, officials, and elected representatives to take a positive position in this area. Climate-related data can expose politically contested ground, and city governments can be 'hesitant to gather or release data that might show their city in a negative light or reveal information that is felt to be proprietary' (Solecki et al. 2015, 126). There is some optimism, however, that ways can be found to develop the requisite data, for example, the C40 does not require cities to verify the data provided in the GPC framework. This may impact on the international agenda in the first instance. However, there is an expectation within the C40 and the climate policy community that such conditions may encourage access to information from cities and city networks. This access should improve the current international situation regarding city governments measurably demonstrating their climate policy's effectiveness (Solecki et al. 2015). Improved data is also a key element in city governments' positions to inform and engage with citizens in the implementation of climate change action plans. Accurate, relevant, and timely performance data is a critical element in the community engagement agenda that strengthens the role of cities in bringing about behaviour change within the community.

Community Engagement Agenda

In democratic systems, policy action ideally requires public input, support, or quiet consent, and implementation requires active engagement of the public. The public governance literature points to trends of increasing collaboration and co-production in the implementation of policy. The need for governments to engage citizens as key stakeholders requires approaches based on evidence and trust. Convincing arguments based on accurate and relevant data can serve as the basis for communicating with citizens on the need to change behaviour. By measurably demonstrating their success through policy evaluation and review, city governments will start to build the requisite evidence and supporting arguments for approaching citizens for behaviour change. Unfortunately, studies that address the question of which factors are most important in changing people's beliefs and attitudes towards climate change are scarce (Hagen et al. 2015). Climate change as a policy issue is highly complex, potentially very pervasive in terms of behaviour change, and for many citizens, it remains an uncertain phenomenon. It is 'generally difficult for people to conceptualise and to relate to their daily activities, arguably because it cannot be easily translated into the language of popular culture' (Lorenzoni and Pidgeon 2006, 74). Policy researchers argue that climate change is the classic 'wicked problem', as one that 'defies resolution because of the enormous interdependencies, uncertainties, circularities, and conflicting stakeholders implicated by any effort to develop a solution' (Rittel and Weber 1973; Head 2008; Lazarus 2008). Policy-making will need to accommodate context and value preferences when deciding options to pursue. Inherent in wicked problems is the lack of clear and definitive solutions. 'You don't so much "solve" a wicked problem as you help stakeholders negotiate shared understanding and shared meaning about the problem and its possible solutions. The objective of the work is coherent action, not final solution' (Conklin 2007, 5). This lack of clear solutions poses a significant challenge for city governments, as climate change becomes a moving feast of new research and changing public attitudes that in combination help shape potential policy options.

Reluctance among residents to change personal activities in order to reduce individual carbon footprint has been among the challenges faced by city governments in implementing climate plans (Wheeler 2008). Communication is an essential means to link scientists, politicians, and the public, and thus can and should play an important and constructive role in enabling public engagement with climate change. Effective communication succeeds in bridging the science–action gap; it does so by fostering public engagement with climate change. This involves cognitive, affective, and behavioural dimensions. That is, people grapple mentally with and gain an understanding of the issue; experience an emotional response, such as interest, concern, or worry; and actively respond by way of changes in climate-relevant behaviour or political action (Lorenzoni and Pidgeon 2006; Moser 2009; Moser and Dilling 2007). Involving residents in climate change policy has become an important challenge for city governments. It has become clear to political decision-makers that 'none of the mitigation and adaptation efforts can succeed without engaging urban residents to support the development or realization of such policies' (Moser 2006, 1).

City governments occupy a unique position as the government closest to the citizens and one with the potential to raise awareness and provide linkages to relevant stakeholders in the implementation of policy. In a 2013 survey of New Yorkers, respondents pointed to the individual and the local government as the two stakeholder groups with the most potential to affect change. Municipally led campaigns for individual behavioural change were seen to be well-suited to New Yorkers, particularly in the context of the city government's own sustainability efforts (NYC

2013). Climate policy holds particular challenges for city governments in terms of both citizens' behaviour change and support for more significant policy initiatives required to 'de-link' the high quality of life from high levels of consumption, identified by Satterthwaite and Dodman (2009), to reduce human contribution to global warming. Unfortunately, the literature has identified poor levels of communication between city governments and residents, which has limited their role in enabling public engagement and support for action (Moser 2006).

Current research points to some lessons for city governments on ways in which they can potentially improve their efforts to encourage residents' participation in climate action plans. Surveys from Spain, the Netherlands, the United Kingdom, and Germany reveal that people are uncertain, indecisive, and lack strong views on the impacts of climate change on future generations (Hagen et al. 2015). With such uncertainty, there are opportunities to promote behaviour change through the well-designed communication programmes that link the individual to global climate change. Critical to the potential success of these programmes will be trust towards the organisation providing climate change information, the perceived level of responsibility for the impact on the climate, and knowledge of the public's risk perception, experiences, and attitudes towards climate change to improve communication programmes and engage the public in a two-way conversation (Hagen et al. 2015). It is clear from the research that people with a better understanding of climate change science and trust in the information provided tend to be more supportive of mitigation efforts (Moser 2010; Spence et al. 2012). Researchers point to the concept of 'carbon capability' as an area where more work needs to be directed (Whitmarsh et al. 2011). There is a need for public understanding of the causes and consequences of carbon emissions, as well as of the ability to reduce emissions. It appears little has been done to consider the meanings of carbon and energy in everyday life and decisions. 'Carbon capability' is the term used in the literature to capture the contextual meanings associated with carbon and individuals' abilities and motivations to reduce emissions.

Some city governments have been active in identifying the most effective ways of influencing citizens' behaviour. Most city governments find that only a limited number of people are willing to do anymore than practice domestic energy conservation; even fewer are prepared to take actions to adapt to climate change. This poses a significant barrier for high levels of public support for climate change policies, which are absolutely necessary to allow policy-makers to develop and implement mitigation and adaptation policies successfully (Hagen et al. 2015). However, there seem to be three major shortcomings of past and current public communication interventions: (1) Most public interventions ought to be but are not designed in an integrative manner, which include communication messages that appeal to cognitive, experiential, and normative dimensions of human behaviour. (2) Current campaigns do not sufficiently target specific behaviours, nor pay sufficient attention to the psychological determinants of the behaviours that they are trying to change. (3) Public campaigns often fail to make the climate change context explicit (Van der Linden 2014).

The City of Sydney attempted to overcome many of these shortcomings in 2009 with the implementation of its *Residential Environment Action Strategy*. Following a community consultation process, consultants established a list of 12 criteria to be applied by the council staff in determining the impact of potential/actual behaviour motivated by sustainability initiatives that echo much of the behaviour change literature. Factors such as targeting the interests and capacities of individuals and communities, and being appropriate to the communities' level of understanding of the problem were included. So too were linking with the communities' understanding of the problem, using a range of approaches, recognising achievement, focusing on building competence, using evaluations as an integral part of the design of the delivery of the programme and of the development of future programmes (Collier and Smith 2009). Another Australian example is the City of Greater Geraldton Region in Western Australia, which adopts a more interactive approach and works closely with Curtin University to improve its sustainability and resilience through engagement of its citizens in collaboration problem-solving and decision-making. Deliberative decision-making and information sharing by the city government revealed that members of the community were far more accepting of sustainability measures and less conservative in their views than what government decision-makers anticipated. On the basis of higher levels of trust based on information exchange, government decision-makers were in a position to implement more comprehensive and ambitious actions on climate change and sustainability than they previously thought possible (Zint and Wolske 2014, 20–21). One of the most significant engagement activities is GreeNYC implemented by the New York Mayor's office in 2010. GreeNYC attempts to increase engagement by residents in the climate change initiatives established in the PlaNYC strategy. In keeping with the approach of the Mayor's office, the programme is essentially a top-down approach that identifies ways to engage residents in already established objectives. GreeNYC tends to follow traditional education and outreach approaches that look for effective ways to improve responses from a target audience (see Box 3.6). The approach taken provides insights into the challenges city governments face in improving the 'carbon capability' of local residents by raising awareness of the meanings of carbon and energy in the everyday lives of New Yorkers.

The need to document and disseminate information about cities' sustainability engagement programmes emerges as a key observation from the case study analysis of successful citizen engagement activity for climate change and sustainability by city governments. As long as engagement strategies are grounded in research-based practices and informed by lessons learned through evaluation, some authors argue that they have the potential to make significant contributions towards fostering the individual, collective, and societal changes needed for sustainability (Zint and Wolske 2014, 26).

We know from the policy literature that for governments to achieve higher levels of engagement and participation by citizens, trust is a key element. To build trust, it is necessary for parties to demonstrate reciprocity and to avoid reneging on undertakings (Head and Alford 2015). But this takes time, and conditions can be difficult to maintain. This is particularly true in a public sector prone to turbulence, and where accountability regimes reduce managers' autonomy and constrain their ability to engage in reciprocity or avoid reneging (ibid.). For residents to engage in a meaningful way, they need to feel that city government authorities use the information acquired through consultative procedures to serve the interests of the community. De Cremer and Tyler (2007)

Box 3.6 GreenNYC, public education for urban environmental behaviour

In 2010, the New York Mayor's Office of Long Term Planning and Sustainability (OLTPS) introduced the GreeNYC initiative as a public education and outreach programme to encourage residents to adopt individual sustainable behaviour to reduce GHG emissions. As part of the GreeNYC programme, the OLTPS conducted research to identify key behaviours that would have the most impact on emissions and help achieve the PlaNYC '30% by 2020' target. The programme assumed that resident compliance (e.g., greater use of new bicycle lanes, more recycling, and increased energy efficiency) was essential to ensure the success of policies introduced by PlaNYC. Research undertaken by the OLTPS identified the 'actions New Yorkers can take, which have the biggest impact on the environment' (NYC 2013). It also identified the motivators that 'drive' behaviour change and established metrics that quantify the environmental impact of the GreeNYC campaigns (ibid.). Utilising the extensive measurement tools established in the PlaNYC project, the research team identified 10 key actions and the corresponding impact on emissions if residents changed behaviour; the results are as follows:

	Action*	Impact, MTCO2e**
1	Replace conventional gas engine car with a gasoline hybrid	1,189,962
2	Switch to electricity produced from non-fossil-fuel or clean sources	859,940
3	Perform a home-energy audit and act on its recommendations	786,346
4	Replace conventional gas engine car with an electric vehicle or plug-in	437,390
5	Weatherize your home	407,929
6	Air dry clothes and use cold water in dishwasher and washing machine	108,724
7	Replace incandescent bulbs with compact fluorescent light bulbs (CFLs)	86,737
8	Eat produce in season and purchase it from farmer's markets, green carts, or coops	50,906
9	In winter, turn down the thermostat 10 degrees when leaving for work and again before bed	41,440
10	Optimize vehicle performance by tuning engine, inflating tires, and using correct motor oil	36,568
	Total	4,005,942
	Percentage of 2011 NYC Inventory***	7.5%

Source: NYC (2013)

Through qualitative research across the city, for example, focus groups and individual interviews, the results revealed that the key driver for behaviour change for New Yorkers is cost savings. This is not surprising in a city where approximately 46% live at or near poverty (Roberts 2014). A key

Box 3.6 (continued)

finding of the research was that residents recognised individual responsibility as a source for change; the authors conclude: 'New Yorkers point to the individual and the local government as the two stakeholder groups with the most potential to affect change' (NYC 2013). The information gathered through the research contributed to the development of GreeNYC campaigns. A series of indicators were developed to assess different campaign methodologies in terms of awareness, progress, and the impact of citizens changing their behaviour. As an example, the campaign targeting the switch to energy-efficient light bulbs (compact fluorescent lamp [CFLs]) produced the following metrics:

Awareness: close to 600 million media impressions were made
Progress: 12% increase in New Yorkers purchasing CFLs, with a 25% increase in sales of CFLs
Impact: estimated carbon impact of the change to CFLs is 56,700 MTCO2e

The GreeNYC research provides important insights into how city governments can identify some of the impacts of their initiatives. It also demonstrates the importance of understanding the residents' views while developing programmes that rely on behaviour change when implementing climate action plans.

propose 'meaningful procedural fairness' as an approach that positively affects cooperation and participation. Such procedural fairness also needs to be 'perceived' by citizens as meaningful, thereby increasing trust and the willingness to be involved.

Performance measurement is essential to building trust and will require strategic approaches with more sophisticated or nuanced methods to programme evaluation (Van Dooren et al. 2015). Typically, this should focus more on the results of the engagement process, because this allows more flexibility concerning the processes by which outcomes are achieved. Evaluation should also recognise two things: (a) the complex feedback loops permeating these processes, and (b) the long lead times often required to address problems associated with climate change, through greater focus on evaluating intermediate and precursor steps (ibid.). Evaluation and communication, and policy learning are fundamental aspects of performance management. City government agencies will require enabling structures and processes to support increased engagement and collaboration in responding to wicked problems generally, and climate change particularly. Meaningful engagement will require more flexible organisational structures, more flexible budgeting and financial systems, and new approaches to human resource management that promote skills more suited to working in open-ended, collaborative, and adaptive situations (Head and Alford 2015). These conclusions lead to the third element of the performance agenda, which focuses on improving governance arrangements within the administration of city governments so that sustainability and climate change issues are embedded considerations within city government policies and management procedures.

Organisational Governance Agenda

The complex and multifaceted nature of climate change policy will require interventions to change established behaviours in the development and implementation of policy across the multiple sectors and responsibilities of city governments. Ongoing discussions are needed on how to embed considerations of climate change into policy processes and institutional systems (Dovers and Herzi 2010). Mainstreaming has emerged as the term to describe the processes in which policy actors have to consider the effects of climate change for their policy domain and decide on implementation measures to reduce their policies' vulnerability to climate effects (Uittenbroek et al. 2013). Mainstreaming has traditionally been regarded as the 'integration of policies and measures to address climate change into ongoing sectoral and development planning and decision making, so as to ensure the long-term sustainability of investments as well as at reduce the sensitivity of development activities to both today's and tomorrow's climate' (Klein et al. 2005, 584). Research on mainstreaming is challenging, as the application of the notion by governments is surprisingly still in its infancy (Wamsler 2015). Frameworks have been developed that illustrate the various dimensions of mainstreaming and the corresponding strategies observed through case study research largely in European contexts—see, for example, Table 3.4.

The literature reveals some of the challenges faced by city governments mainstreaming climate change into the policy process. In practice, there is a need for constant reconsideration of the necessity for

Table 3.4 Dimensions and strategies for mainstreaming climate change

Dimensions	Strategies
Horizontal mainstreaming	
1. Add-on	Short-term projects that are not part of the organisational unit's core work
2. Programmatic	Modifications of the organisational unit's work by integrating elements of climate change into projects
3. Inter and intra organisational	Promotion of collaboration between the organisational unit and the stakeholders, leading to joint actions
Vertical mainstreaming	
1. Managerial	Modification of managerial and working structures to institutionalise climate change
2. Regulatory	Modification of planning procedures and related activities, leading to integration of climate change
3. Directed	Support of or redirection of focus onto aspects related to climate change

Source: Adapted from Wamsler (2015)

mainstreaming as a result of the barriers and opportunities from other policy areas and from the policy process itself (Uittenbroek et al. 2013). In many respects, there is going to be an inbuilt conflict between many of the functions of city governments, for example, urban planning and economic development, and the objectives of sustainability. Much of the research has been directed to the mainstreaming of adaptation to the impacts of climate change on cities (Rauken et al. 2015; Wamsler 2015). City governments are at the forefront of these impacts, for example, extreme weather events such as flooding and extreme temperature causing damage to infrastructure. This reality results in city governments being the most obvious focus for the development and implementation of adaptation policies to prepare and protect citizens from these impacts.

City government policy actors need to apply strategies that focus on establishing synergies across the different policy areas and making deliberate decisions that take into account current and future impacts. Canadian research reveals some of the critical issues involved in establishing the conditions within the organisational culture to promote effective implementation of climate policy. One of the critical enablers of action is the institutionalisation of climate measures across the relevant policy responsibilities of local governments (Burch 2010). The planning literature argues that adaptation is an area of particular relevance to city governments, as the 'more adaptation is integrated in functionally linked policy documents and processes, the better the chance for society to become climate proof' (Uittenbroek et al. 2013, 400). Researchers express concern about the capacity and/or willingness of city governments to pursue an agenda of mainstreaming climate considerations across their policy sectors. There is recognition that mainstreaming requires interventions that will lead to behaviour change across organisational units within city governments. Detailed discussions on exactly how to redesign processes and institutions are beginning to emerge; however, present indications suggest that change will be slow, as there are entrenched interests that resist behaviour change. In many instances, the changes required will be considerable, as they imply a fundamental reworking of institutional structures, organisational culture, and policy procedures with government organisations in order to improve policy implementation (Burch 2010).

Mainstreaming is an internal governance issue with city governments learning to embed sustainability into policy and planning processes. Much of the evidence emerges from Europe, with case studies that provide policy-makers with some insights into the challenges faced by city governments when mainstreaming adaptation into their management and policy processes (Wamsler 2015; Rauken et al. 2015; Uittenbroek et al. 2013). Establishing an organisational culture that supports innovation and collaboration appears to be the fundamental insight, as the issues raised by climate change require new ways of examining long-term problems faced by city governments. Moser and Ekstrom (2010) established a comprehensive framework for identifying the barriers within the policy process to mainstreaming adaptation. These barriers serve as key areas of focus for city governments as they develop their climate change action plans. The framework serves as a useful three-stage heuristic where barriers and opportunities arise: *understanding* (in general social, cognitive, and organisational/institutional), *planning* (financial, technological, and organisational/institutional), and *management* (financial and organisational/institutional) (see Table 3.5).

Table 3.5 Barriers and opportunities for mainstreaming

Understanding	Planning	Management
Problem data and awareness raising	Development of adaptation options	Implementation of selected options
Information gathering	Assessment of options	Monitoring of the environment and the outcome of the realised option
Problem definition	Selection of options	Evaluation to support institutional and social learning

Source: Adapted from Moser and Ekstrom (2010)

Some of the barriers and opportunities crosscut throughout the process and might reoccur in a later phase. Additionally, the policy process is iterative and not linear. This implies that barriers and opportunities from an earlier phase could repeat themselves as the policy process progresses (Uittenbroek et al. 2013).

Case study research from five Norwegian cities reveals that where adaptation received political attention, researchers found more attempts at mainstreaming (Rauken et al. 2015; Juhola and Westerhoff 2011). Policy development is slower in city governments that pursue a horizontal, cross-sectoral approach to mainstreaming that encourages cooperation and coordination between organisational units. However, in city governments that pursue a vertical sector approach, with top-down directives from the chief executive and/or political leaders, approaches to mainstreaming are better maintained (Rauken et al. 2015, 419). The main issue with the horizontal approach is that it may prove insufficient in the long term, as climate initiatives might not be pursued with equal enthusiasm in all sections of the organisation. Political attention can be the prime initiator of a mainstreamed approach and raises its profile as a priority for organisational units in a city government. Canadian research confirms this view, whereby 'stronger leadership may both drive innovation and assure technical staff that climate change priorities will not suddenly evaporate in the face of other pressing concerns' (Burch 2010, 296). It seems clear from the case study evidence that for a more long-lasting and consistent policy, the cooperation demanded from a horizontal, cross-sectoral approach may have advantages, as the approach has the potential to find holistic solutions to a crosscutting issue. Such a strategic approach supports the multiple objectives that are necessary when attempting to establish consistency across the different policy areas of city governments. The implementation of climate change plans will require cooperation and collaboration in areas of key city responsibility. These areas include planning to reduce the impact of sprawl, promoting alternative forms of commuting, and supporting energy efficiency in public and private buildings. Research demonstrates that embedding broader sustainability goals in long-range strategic planning contributes to the ongoing efforts to adapt institutions to the complex and uncertain futures associated with a changing climate (Burch 2010). The establishment of strategic plans endorsed by the political leadership, which incorporate climate change and sustainability considerations, to provide the consistency needed for the benefits of both the vertical and horizontal approaches to mainstreaming.

The Promise of Performance Management

A performance management–based approach holds significant potential in contributing to the three agendas for sustainability and climate change by city governments. Performance supports an approach that assesses achievements against the original objectives established within the climate action plan. Through these assessments, city governments develop transparent decision-making processes and strengthen accountability mechanisms against their policy actions. A performance approach to management initially examines whether a 'plan or objective is included and affects future decision making' (Uittenbroek et al. 2013, 402). Indicators and measures contribute to improved performance management within city governments in order to help coordinate a more informed set of planning norms and practices, more effective infrastructure investment and urban management, and a more inclusive urban governance (McCarney 2012). Through a process of evaluation and review, the aim to mainstream climate-based objectives within the activities of each organisational unit can be examined. Individual units can be called to account for their mainstreaming efforts. Actors within these units will be responsible for policy implementation processes. Responsibility will include understanding what is feasible in terms of financial, social, and technical constraints (Uittenbroek et al. 2013). Such evaluations form part of the policy learning process. A key aspect of performance management is that it is designed to identify what works and why. This knowledge contributes to both policy and organisational learning that can shape approaches to behaviour change that improve the opportunities to achieve climate change policy objectives (Fig. 3.3).

Performance management also provides opportunities to review objectives as they are achieved and to consider an alternative course of action. Performance reporting provides opportunities for social learning, as the city government provides public reports on its actions to mitigate and adapt to the impacts of climate change. Importantly, for policy-makers, there appears to be a link between performance reporting and trust. If citizens find the performance information from government agencies to be credible, they are more likely to believe what is being reported (James and Van Ryzin 2017). How measures are made, results incorporated, and findings used are all heavily impacted by contextual considerations. Surveys of US cities reveal that even 'exemplary jurisdictions face challenges using evidence-based, data-driven performance management approaches' (Sanger 2008, s70). Promoting best practice requires better scholarship to understand the determinants of successful implementa-

1	Capacity to assess policy achievements against original objectives
2	Evidence to support coordination of more informed planning norms and practices
3	Provide evidence for more effective infrastructure investment and urban management
4	Evaluation and review to support mainstreaming of climate objectives across organisational units
5	Improved understanding on what is feasible in terms of financial, social and technical constraints
6	Evidence for improved policy learning
7	Review of policy objectives as they are achieved / or not, consider alternative courses of action
8	Opportunities of social learning, transparency and accountability through public reports and climate action
9	Build community trust through credible information
10	Evidence to support best practice methods

Fig. 3.3 Promise of performance management and the implementation of climate policy by city governments

tion, adoption, and use of performance measurement. The outcomes of performance-based systems have generally been unmeasured and little is known about their cost-effectiveness or endurance over time (Gerrish 2016). The promise is there, but our expectations should be tempered (Sanger 2008). The act of measuring performance may not improve performance, but managing performance might. Performance management systems require the application of fundamental elements, including clarity of goals, relevant measures, and information that is used for management purposes (Ammons 2015). The next chapter will outline a framework for examining performance management in city governments that has the potential to meet the requirements of the three agendas outlined in this chapter. The framework provides the theoretical basis for the analysis of city governments' performance schemes in the remaining chapters of the book.

References

ACC (Adelaide City Council). (2011). *Energy Management Action Plan*. South Australia: Adelaide.

ACC (Adelaide City Council). (2014). *City of Adelaide Community Reduces Carbon Emissions by 19%*. Adelaide City Council Media Release, October 21.

Ammons, D. N. (2013). Signs of Performance Measurement Progress Among Prominent City Governments. *Public Performance & Management Review, 36*(4), 507–528.

Ammons, D. (2015). Getting Real About Performance Management: Using Performance Information to Improve Services is Key. *Public Management*. December. ICMA Publications, PM Magazine. http://icma.org/en/Article/106489/Getting_Real_About_Performance_Management

BCC (Brisbane City Council). (2015). *Annual Report*. Queensland: Brisbane.

Betsill, M., & Bulkeley, H. (2007). Looking Back and Thinking Ahead: A Decade of Cities and Climate Change Research. *Local Environment, 12*(5), 447–456.

Blackhurst, M., Matthews, H. S., Sharrard, A. L., Hendrickson, C. T., & Azevedo, I. L. (2011). Preparing US Community Greenhouse Gas Inventories for Climate Action Plans. *Environmental Research Letters, 6*(3), 034003.

Brander, M., Carstairs, S., & Topp, C. F. (2013). Global protocol for Community Scale Greenhouse Gas Emissions: A Trial Application in the West Highlands of Scotland. *Greenhouse Gas Measurement and Management, 3*(3–4), 149–165.

Bulkeley, H., & Betsill, M. M. (2003). *Cities and Climate Change. Urban Sustainability and Global Environmental Governance.* London: Routledge.

Bulkeley, H., & Betsill, M. (2005). Rethinking Sustainable Cities: Multilevel Governance and the 'Urban' Politics of Climate Change. *Environmental Politics, 14*(1), 42–63. DOI 10.1080/0964401042000310178.

Bulkeley, H., & Betsill, M. M. (2013). Revisiting the Urban Politics of Climate Change. *Environmental Politics, 22*(1), 136–154.

Burch, S. (2010). Transforming Barriers into Enablers of Action on Climate Change: Insights from Three Municipal Case Studies in British Columbia, Canada. *Global Environmental Change, 20*(2), 287–297.

Butler, T. M., Lawrence, M. G., Gurjar, B. R., Van Aardenne, J., Schultz, M., & Lelieveld, J. (2008). The Representation of Emissions from Megacities in Global Emission Inventories. *Atmospheric Environment, 42*(4), 703–719.

C40. (2014). *Global Protocol for Community Scale Greenhouse Gas Emission Inventories.* http://www.ghgprotocol.org/city-accounting

C40. (2016). *Unlocking Climate Action in Megacities.* http://www.c40.org/researches/unlocking-climate-action-in-megacities

City of Stockholm. (2012). The Stockholm Environment Programme 2012–2015. http://international.stockholm.se/globalassets/ovriga-bilderoch-filer/the-stockholm-environment-programme-2012-2015.pdf

Collier, G., & Smith, P. (2009). Beyond Lip Service: A Council Approach to Planning for Behaviour Change. *Australian Journal of Environmental Education, 25*, 129–138.

Conklin, J. (2007). Rethinking Wicked Problems. *NextD Journal, 10*, 1–30. http://nextd.org

Croci, E., Melandri, S., & Molteni, T. (2011). Determinants of Cities' GHG Emissions: A Comparison of Seven Global Cities. *International Journal of Climate Change Strategies and Management, 3*(3), 275–300. DOI 10.1108/17568691111153429.

D'Avignon, A., Carloni, F. A., La Rovere, E. L., & Dubeux, C. B. S. (2010). Emission Inventory: An Urban Public Policy Instrument and Benchmark. *Energy Policy, 38*(9), 4838–4847.

De Cremer, D., & Tyler, T. R. (2007). The Effects of Trust in Authority and Procedural Fairness on Cooperation. *Journal of Applied Psychology, 92*(3), 639.

Dietz, T., Dan, A., & Shwom, R. (2007). Support for Climate Change Policy: Social Psychological and Social Structural Influences. *Rural Sociology, 72*(2), 185–214. DOI 10.1526/003601107781170026.

Dodman, D. (2009). Blaming Cities for Climate Change? An Analysis of Urban Greenhouse Gas Emissions Inventories. *Environment and Urbanization, 21*(1), 185–201.

Doucet, C. (2007). *Urban Meltdown: Cities, Climate Change and Politics as Usual.* Gabriola Island, BC: New Society Publishers.

Dovers, S. R., & Hezri, A. A. (2010). Institutions and Policy Processes: The Means to the Ends of Adaptation. *Wiley Interdisciplinary Reviews: Climate Change, 1*(2), 212–231.

Duren, R. M., & Miller, C. E. (2011). Towards Robust Global Greenhouse Gas Monitoring. *Greenhouse Gas Measurement & Management, 1*(2), 80–84.

Einstein, K., Glick, D., & Lusk, K. (2014). *Mayoral Policy Making: Results from the 21st Century Mayor's Leadership Survey.* Boston: Boston University Initiatives on Cities.

Engel, K. (2009). Whither Subnational Climate Change Initiatives in the Wake of Federal Climate Legislation. *Publius: Journal of Federalism, 39*(3), 432–454.

Erickson, P., Lazarus, M., Chandler, C., & Schultz, S. (2013). Technologies, Policies and Measures for GHG Abatement at the Urban Scale. *Greenhouse Gas Measurement and Management, 3*(1-2), 37–54. DOI 10.1080/20430779.2013.806866.

Eurobarometer. (2009). Europeans' Attitudes Towards Climate Change. Special Eurobarameter 313. European Parliament/European Commission. Retrieved from http://ec.euro-pa.eu/public_opinion/archives/ebs/ebs_313_en.pdf

Fielding, K., Head, B., Laffan, W., Western, M., & Hoegh-Guldberg, O. (2012). Australian Politicians' Beliefs About Climate Change: Political Partisanship and Political Ideology. *Environmental Politics, 21*(5), 712–733. DOI 10.1080/09644016.2012.698887.

Garren, S., & Brinkmann, R. (2012). Linking Greenhouse Gas Emissions to Greenhouse Gas Reduction Policies at the Local Government Level in Florida. *Greenhouse Gas Measurement and Management, 2*(2-3), 146–161.

Gerrish, E. (2016). The Impact of Performance Management on Performance in Public Organizations: A Meta-Analysis. *Public Administration Review, 76*(1), 48–66.

Hagen, B., Middel, A., & Pijawka, D. (2015). European Climate Change Perceptions: Public Support for Mitigation and Adaptation Policies. *Environmental Policy and Governance.* DOI 10.1002/eet.1701.

Harvey, F. 2016. Four of the World's Biggest Cities to Ban Diesel Cars from Their Centres, *The Guardian*, December 2. https://www.theguardian.com/environment/2016/dec/02/four-of-worlds-biggest-cities-to-ban-diesel-cars-from-their-centres

Hatry, H. P. (2006). *Performance Measurement: Getting Results*. Washington, DC: The Urban Institute.

Head, B.W. (2008). Wicked Problems in Public Policy. *Public Policy, 3*(2).

Head, B. W., & Alford, J. (2015). Wicked Problems: Implications for Public Policy and Management. *Administration & Society, 47*(6), 711–739.

Hsu, A., Cheng, Y., Weinfurter, A., Xu, K., & Yick, C. (2016). Track Climate Pledges of Cities and Companies. *Nature, 532*, 303–306. http://www.nature.com/news/track-climate-pledges-of-cities-and-companies-1.19764

IPCC (International Panel on Climate Change). (2014). Summary for Policymakers. In O. Edenhofer, R. Pichs-Madruga, Y. Sokona, E. Farahani, S. Kadner, K. Seyboth, A. Adler, I. Baum, S. Brunner, P. Eickemeier, B. Kriemann, J. Savolainen, S. Schlmer, C. von Stechow, T. Zwickel, & J. C. Minx (Eds.), *Climate Change 2014, Mitigation of Climate Change. Contribution of Working Group III to the Fifth Assessment Report of the Intergovernmental Panel on Climate Change*. Cambridge: Cambridge University Press.

James, O., & Van Ryzin, G. G. (2017). Incredibly Good Performance: An Experimental Study of Source and Level Effects on the Credibility of Government. *The American Review of Public Administration, 47*(1), 23–35.

Jones, S. (2012). A Tale of Two Cities: Climate Change Policies in Vancouver and Melbourne: Barometers of Cooperative Federalism? *International Journal of Urban and Regional Research, 36*(6), 1242–1267.

Juhola, S., & Westerhoff, L. (2011). Challenges of Adaptation to Climate Change Across Multiple Scales: A Case Study of Network Governance in Two European Countries. *Environmental Science & Policy, 14*(3), 239–247.

Kennedy, C., Demoullin, S., & Mohareb, E. (2012). Cities Reducing Their Greenhouse Gas Emissions. *Energy Policy, 49*, 774–777.

Kern, K., & Alber, G. (2009). Governing Climate Change in Cities: Modes of Urban Climate Governance in Multi-Level Systems, in Competitive Cities and Climate Change: OECD Conference Proceedings, 9–10 October 2008, Milan, Italy http://www.oecd.org/dataoecd/12/38/42554913.pdf

Kingdon, J. (2003). *Agendas, Alternatives, and Public Policies* (2nd ed.). New York, NY: Longman.

Kitchin, R., Lauriault, T. P., & McArdle, G. (2015). Knowing and Governing Cities Through Urban Indicators, City Benchmarking and Real-Time Dashboards. *Regional Studies, Regional Science, 2*(1), 6–28.

Klein, R. J., Schipper, E. L. F., & Dessai, S. (2005). Integrating Mitigation and Adaptation into Climate and Development Policy: Three Research Questions. *Environmental Science & Policy, 8*(6), 579–588.

Kousky, C., & Schneider, S. H. (2003). Global Climate Policy: Will Cities Lead the Way? *Climate Policy, 3*(4), 359–372.

Krause, R. M. (2011). Policy Innovation, Intergovernmental Relations, and the Adoption of Climate Protection Initiatives by US Cities. *Journal of Urban Affairs, 33*(1), 45–60.

Kroll, A. (2013). The Other Type of Performance Information: Non Routine Feedback, Its Relevance and Use. *Public Administration Review, 73*(2), 265–276.

Lazarus, R. J. (2008). Super Wicked Problems and Climate Change: Restraining the Present to Liberate the Future. *Cornell Law Review, 94*, 1153–1234.

Leccese, M., & McCormick, K. (2000). *Charter of the New Urbanism*. Congress for the New Urbanism. New York: McGraw Hill.

Lee, T., & Painter, M. (2015). Comprehensive Local Climate Policy: The Role of Urban Governance. *Urban Climate, 14*, 566–577.

Lorenzoni, I., & Pidgeon, N. F. (2006). Public Views on Climate Change: European and USA Perspectives. *Climatic Change, 77*(1), 73–95.

Mai, Q., & Francesch-Huidobro, M. (2014). *Climate Change Governance in Chinese Cities*. London: Routledge.

Marcotullio, P. J., Sarzynski, A., Albrecht, J., Schulz, N., & Garcia, J. (2013). The Geography of Global Urban Greenhouse Gas Emissions: An Exploratory Analysis. *Climatic Change, 121*(4), 621–634.

Marr, B. (2009). *Managing and Delivering Performance*. London: Elsevier.

McCarney, P. L. (2012). City Indicators on Climate Change: Implications for Governance. *Environment and Urbanization Asia, 3*(1), 1–39.

McCarney, P. (2013). Cities and Governance: Coming to Terms with Climate Challenges. In *Climate Change Governance* (pp. 85–103). Heidelberg: Springer.

McConnell, A. (2010). Policy Success, Policy Failure and Grey Areas In-between. *Journal of Public Policy, 30*(03), 345–362.

McCright, A. M., & Dunlap, R. E. (2011). The Politicization of Climate Change and Polarization in the American Public's Views of Global Warming, 2001–2010. *The Sociological Quarterly, 52*(2), 155–194. DOI 10.1111/j.1533-8525.2011.01198.x.

Millard-Ball, A. (2013). The Limits to Planning: Causal Impacts of City Climate Action Plans. *Journal of Planning Education and Research, 33*(1), 5–19.

Miller, N., Cavens, D., Condon, P., & Kellet, R. (2009). Policy, Urban Form, and Tools for Measuring and Managing Greenhouse Gas Emissions: The North American Problem. *University of Colorado Law Review, 80*, 977.

Moser, S. C. (2006). Talk of the City: Engaging Urbanites on Climate Change. *Environmental Research Letters, 1*(1).

Moser, S. C. (2009). Communicating Climate Change and Motivating Civic Action: Renewing, Activating, and Building Democracies. In *Changing Climates in North American Politics: Institutions, Policymaking and Multilevel Governance* (pp. 283–302). Cambridge: Cambridge University Press.

Moser, S. C. (2010). Communicating Climate Change: History, Challenges, Process and Future Directions. *Wiley Interdisciplinary Reviews: Climate Change, 1*(1), 31–53.

Moser, S. C., & Dilling, L. (2007). Toward the Social Tipping Point: Creating a Climate for Change. In *Creating a Climate for Change: Communicating Climate Change and Facilitating Social Change* (pp. 491–516). Cambridge: Cambridge University Press.

Moser, S. C., & Ekstrom, J. A. (2010). A Framework to Diagnose Barriers to Climate Change Adaptation. *Proceedings of the National Academy of Sciences, 107*(51), 22026–22031.

Moynihan, D. P., Fernandez, S., Kim, S., LeRoux, K. M., Piotrowski, S. J., Wright, B. E., & Yang, K. (2011). Performance Regimes Amidst Governance Complexity. *Journal of Public Administration Research and Theory, 21*(suppl 1), i141–i155.

Nelsen, A. (2017). European Commission Issues Final Warning to UK Over Air Pollution Breaches. *The Gaurdian*. https://www.theguardian.com/environment/2017/feb/15/european-commission-issues-final-warning-to-uk-over-air-pollution-breaches

NYC (City of New York). (2013). *New Yorkers' Contributions to a Greener, Greater City: A Behavioural Impact Study*. Mayor's Office of Long Term Planning and Sustainability. New York.

OECD. (2009). *Measuring Government Activity*. Paris: OECD.

Rauken, T., Mydske, P. K., & Winsvold, M. (2015). Mainstreaming Climate Change Adaptation at the Local Level. *Local Environment, 20*(4), 408–423.

Reams, M. A., Clinton, K. W., & Lam, N. S. (2012). Achievement of Climate Planning Objectives Among US Member Cities of the International Council for Local Environmental Initiatives (ICLEI). *Low Carbon Economy, 3*(4).

Reckien, D., Flacke, J., Dawson, R. J., Heidrich, O., Olazabal, M., Foley, A., Hamann, J. P., Orru, H., Salvia, M., Hurtado, S. D. G., & Geneletti, D. (2014). Climate Change Response in Europe: What's the reality? Analysis of

Adaptation and Mitigation Plans from 200 Urban Areas in 11 Countries. *Climatic Change, 122*(1-2), 331–340.
Revi, A. (2008). Climate Change Risk: An Adaptation and Mitigation Agenda for Indian Cities. *Environment and Urbanization, 20*(1), 207–229.
Rittel, H. W., & Webber, M. M. (1973). Dilemmas in a General Theory of Planning. *Policy Sciences, 4*(2), 155–169.
Roberts, S. (2014). Nearly Half of New Yorkers Are Struggling to Get by, Study Finds. *New York Times*, April 30.
Robinson, P. J., & Gore, C. D. (2005). Barriers to Canadian Municipal Response to Climate Change. *Canadian Journal of Urban Research, 14*(1), 102.
Román, M. (2010). Governing from the Middle: The C40 Cities Leadership Group. *Corporate Governance: The International Journal of Business in Society, 10*(1), 73–84.
Sánchez-Rodríguez, R., Seto, K. C., Solecki, W. D., Kraas, F., & Laumann, G. (2005). Science Plan: Urbanization and Global Environmental Change. In *Science Plan: Urbanization and Global Environmental Change*. Bonn, Germany: IDHP.
Sanderson, I. (2002). Evaluation, Policy Learning and Evidence-Based Policy Making. *Public Administration, 80*(1), 1–22.
Sanger, M. B. (2008). From Measurement to Management: Breaking Through the Barriers to State and Local Performance. *Public Administration Review, 68*(s1).
Satterthwaite, D. (2008). Cities' Contribution to Global Warming: Notes on the Allocation of Greenhouse Gas Emissions. *Environment and Urbanization, 20*(2), 539–549.
Satterthwaite, D., & Dodman, D. (2009). The Role of Cities in Climate Change. *State of the World*, 75–77.
Scanu, E., & Cloutier, G. (2015). Why Do Cities Get Involved in Climate Governance? Insights from Canada and Italy. *Environnement Urbain (Urban Environment), 9*, 2–19.
Seto, K. C., Dhakal, S., Bigio, A., Blanco, H., Delgado, G. C., Dewar, D., Huang, L., Inaba, A., Kansal, A., Lwasa, S., McMahon, J. E., Müller, D. B., Murakami, J., Nagendra, H., & Ramaswami, A. (2014). Human Settlements, Infrastructure and Spatial Planning. In O. Edenhofer, R. Pichs-Madruga, Y. Sokona, E. Farahani, S. Kadner, K. Seyboth, A. Adler, I. Baum, S. Brunner, P. Eickemeier, B. Kriemann, J. Savolainen, S. Schlömer, C. von Stechow, T. Zwickel, & J. C. Minx (Eds.), *Climate Change 2014: Mitigation of Climate Change. Contribution of Working Group III to the Fifth Assessment Report of the Intergovernmental Panel on Climate Change*. Cambridge: Cambridge University Press.

Sharp, E. B., Daley, D. M., & Lynch, M. S. (2011). Understanding Local Adoption and Implementation of Climate Change Mitigation Policy. *Urban Affairs Review, 47*(3), 433–457.

Sippel, M. (2011). Urban GHG Inventories, Target Setting and Mitigation Achievements: How German Cities Fail to Outperform Their Country. *Greenhouse Gas Measurement and Management, 1*(1), 55–63.

Sippel, M., & Jenssen, T. (2009). *What About Local Climate Governance? A Review of Promise and Problems.* MPRA Paper No. 20987, posted 26. February 2010.

Solecki, W., Seto, K. C., Balk, D., Bigio, A., Boone, C. G., Creutzig, F., Fragkias, M., Lwasa, S., Marcotullio, P., Romero-Lankao, P., & Zwickel, T. (2015). A Conceptual Framework for an Urban Areas Typology to Integrate Climate Change Mitigation and Adaptation. *Urban Climate, 14*, 116–137.

Spence, A., Poortinga, W., & Pidgeon, N. (2012). The Psychological Distance of Climate Change. *Risk Analysis, 32*(6), 957–972.

The Economist. (2016). Particular About Particulates, https://www.economist.com/news/asia/21688447-bold-experiment-has-improved-delhisair-indians-want-more-particular-about-particulates?zid=306&ah=1b164dbd43b0cb27ba0d4c3b12a5e227

Uittenbroek, C. J., Janssen-Jansen, L. B., & Runhaar, H. A. (2013). Mainstreaming Climate Adaptation into Urban Planning: Overcoming Barriers, Seizing Opportunities and Evaluating the Results in Two Dutch Case Studies. *Regional Environmental Change, 13*(2), 399–411.

UN. (1992). *United Nations Framework Convention on Climate Change.* https://unfccc.int/resource/docs/convkp/conveng.pdf

Van der Linden, S. (2014). Towards a New Model for Communicating Climate Change. In S. A. Cohen, J. E. S. Higham, P. Peeters, & S. Gossling (Eds.), *Understanding and Governing Sustainable Tourism Mobility: Psychological and Behavioural Approaches.* London: Routledge.

Van Dooren, W., Bouckaert, G., & Halligan, J. (2015). *Performance Management in the Public Sector.* London: Routledge.

Wamsler, C. (2015). Mainstreaming Ecosystem-Based Adaptation: Transformation Toward Sustainability in Urban Governance and Planning. *Ecology and Society, 20*(2).

Wheeler, S. M. (2008). State and Municipal Climate Change Plans: The First Generation. *Journal of the American Planning Association, 74*(4), 481–496.

Whitmarsh, L., Seyfang, G., & O'Neill, S. (2011). Public Engagement with Carbon and Climate Change: To What Extent is the Public 'Carbon Capable'? *Global Environmental Change, 21*(1), 56–65.

Zerbinati, S., & Souitaris, V. (2005). Entrepreneurship in the Public Sector: A Framework of Analysis in European Local Governments. *Entrepreneurship & Regional Development, 17*(1), 43–64.

Zint, M., & Wolske, K. S. (2014). From Information Provision to Participatory Deliberation: Engaging Residents in the Transition Toward Sustainable Cities. In *The Elgar Companion to Sustainable Cities: Strategies, Methods and Outlook* (pp. 188–209). Northampton, MA: Edward Elgar Publishing.

4

Performance Management: An Analytical Framework

Introduction

City governments have been subject to the forces of reform experienced by all levels of government in the latter part of the twentieth century. Drivers of change included community demands for a wider range of services, increased focus on cost-effective service delivery, and the adoption of private sector managerial processes to improve efficiencies through management and planning. Reforms based on planning, programming, and budgeting (PPB) have been popular in US local governments since the late 1960s. The PPB approach promoted the establishment of goals and strategic planning with resource allocation along programme lines and financial practices based on a cost–benefit analysis (Schultze 1968). Around the same time, centuries of tradition and patronage in UK local governments were subject to a review by the high-level Redcliffe Maude Royal Commission. Part of the Commission's investigation identified the need to break down internal departmental barriers within local government administration that had been based on benefiting local interests at the expense of the broader public interest. The recommendations of the Commission included the promotion of rational 'whole of government' approaches to

S. Jones, *Cities Responding to Climate Change*,
DOI 10.1007/978-3-319-64810-1_4

help ensure council programmes were blended and coordinated (Bains 1972). The recommendations were supported by the principles of planning, effective resource allocation, and organisational performance directed towards policy outcomes. These changes established functional and structural transition in local governments in many countries that included a shift from administration to management supported by assumptions of the benefits corporate rationality (Tucker 1997).

Since the 1990s, a broad set of neoliberal ideals under the label of New Public Management (NPM) promoted the transformation of classic rule-based government bureaucracies to adopt private sector management practices and become responsive, innovative, high-performing organisations that would respond to government direction and achieve outcomes for communities (Osborne and Gaebler 1992; Heinrich 2002). The emergence of *strategic urbanism* (Bulkeley and Betsill 2013), an approach adopted by city governments to change many of their modes of governance and work with the private sector in tackling climate change issues, has a direct link to NPM principles. Performance management practices established in the private sector were promoted in the public management literature and focused on reforms to government bureaucracies that would initiate a reinvention of management practices to achieve results. The OECD supported performance management as a means to help improve governance and transparency by enabling governments to quantify promises, and measure and report their actions in ways that would allow citizens, managers, and politicians to make meaningful decisions about increasingly complex government activities (OECD 2009). The performance management perspective examines performance-related questions about organisational governance processes and procedures; it also has the potential to promote a broader level of analysis that could help reveal attributes for developing, implementing, and evaluating climate policies by city governments. The conclusions of the previous chapter raised the potential of performance management to assist city governments facing the policy and governance challenges when responding to climate change. This chapter will provide an in-depth analysis of performance management and develop a framework to compare the performance management systems in the case study chapters of leading cities implementing climate action plans. This analysis will help identify potential lessons for other cities responding to climate change to be outlined in the concluding chapter.

Performance Management Doctrine

The performance management doctrine advocates the reduction in bureaucratic rule-based constraints and the accountability of public sector managers for their results (Behn 2002). According to the doctrine, accountability and effectiveness in government agencies are achieved through the setting of targets and the measuring and monitoring of performance. There are two key mechanisms for improving performance. The first is the application of incentives and sanctions for managers to improve performance. The second is policy learning, which should take place when initiated by the performance information made available through measurement and reporting. The key premise of the performance doctrine is that improvements in performance are based on increased autonomy of the agency and its managers (Behn 2002; Jakobsen and Mortensen 2016). Research shows that innovation in public organisations is largely driven by middle- and lower-level managers working in organisational cultures that support learning and knowledge sharing (Gahan et al. 2016). Without the autonomy, supported by fewer bureaucratic rules, there is a risk that managers will be constrained in their potential to improve performance (Moynihan 2009). This trade-off means that important operating decisions can be made at the organisational unit level and performance information is forwarded to senior managers and political representatives, providing assurance that objectives are being achieved (Ammons 2015). Current research supports the view that performance can be improved when there is greater management autonomy to establish performance-based processes and procedures at the unit level (Hvidman and Andersen 2014; Jakobsen and Mortensen 2016). Performance advocates argue that good-quality data is the basis of information that can be used to answer key governance questions relating to better allocation of resources, decisions about strategy, re-engineering processes, staff motivation, and accountability (Moynihan 2009). The promise is that measurement—the quantitative representation of the quality or quantity of inputs, outputs, and/or outcomes (Sterck and Scheers 2006)—can offer valuable information for policy-makers and practitioners by assisting with comparison, prediction, and evaluation.

A performance management system requires an explicit policy on measurement and managing the different functions and their performances. For the majority of city governments, the development of accurate and relevant performance measurement protocols for climate change continues to be a source of debate and experimentation. The underlying assumption is that the introduction of performance information will ultimately benefit government actions, as it leads to purposeful use by public officials to manage their programmes or allocate resources in ways that improve performance (Moynihan et al. 2011). An important finding in the research is that performance management systems are often partially adopted (Pollitt and Bouckaert 2004). Partial adoption means that there has been some formal adoption of a performance information system but this has not been accompanied by an increase in managerial authority (Moynihan 2009). As a result, managers are faced with increased demands to improve performance without the requisite authority to change management processes and procedures. In many instances, public sector managers remain constrained by the rule-based culture but face increased demands to improve performance (Christensen and Lægreid 2010). The performance management literature identifies essential elements of various schools of thought that contributed to the promotion and application of performance as a management approach in the public sector (Ammons and Roenigk 2015). In examining the application and engagement of city governments with performance management, these elements will provide guideposts for analysis. Table 4.1 summarises the essentials of the performance doctrine.

Table 4.1 Performance management doctrine

Organisational goals need to be clear and specific.
Performance information is available, relevant, and actionable.
Results focus on outcomes not inputs and compliance.
Senior executives are engaged in priority setting, monitoring of performance reports, data-influenced decision-making, monitoring of department performance, and frequent feedback.
Devolved decision-making is involved, with operational decisions made at departmental or programme level.
Managerial flexibility exists in the use of financial and human resources, and there is accountability for results.
Incentives and sanctions are tied to performance results.

Source: Adapted from Ammons and Roenigk (2015)

The following sections of the chapter focus on four elements that encapsulate the fundamentals of the performance doctrine. These elements also provide a basis for establishing the analytical tool to be used to examine levels of engagement with performance management in city governments. In terms of climate policy action by city governments, we need to analyse their performance management systems according to

1. the *clarity* of the goals to be pursued by organisational units leading to the achievement of the climate change outcomes outlined in a strategic plan;
2. the collection and processing of *measurement* data into performance-relevant information pertaining to climate change policy objectives;
3. the *incorporation* of performance management relevant to climate change objectives into the governance processes and procedures within the city government;
4. the *use* of performance information to support policy learning and behaviour change.

Clarity of Goals

To measurably demonstrate their success in climate policy, city governments first need to be clear on what they are setting out to achieve. As we have seen in Chap. 3, clarity is an important issue because one of the central weaknesses of many city governments' climate initiatives is the setting of ambitious goals that are beyond their capacity to achieve. Goal setting by public sector agencies has been an important area of research in the public policy literature, and goal clarity is a fundamental aspect of an effective performance management system. Managers in city governments need to be aware of the goals, priorities, and expectations in the context of climate change policy. Chapter 3 provided some insights into the presence of climate change plans within some city governments. Many cities have adopted a strategic management approach that emphasises the linkages between organisational units in the implementation of climate policy. Linkages across the organisation contribute to the clarity needed for managers of the different operational functions to be aware of their contribution to the overall objectives of the city government in implementing a climate change plan. The inclusion of a performance component to the plan requires the setting of targets and the development of indicators to provide clarity of goals. The performance management doctrine stresses the importance of this clarity as critical in informing decisions, monitoring progress, and identifying opportunities for improvement in quality and efficiency (Ammons and Roenigk 2015; Moynihan and Pandey 2004).

Strategic planning has been a feature of local government goal setting in many countries since the late 1980s (Gordon 2013). A strategic plan is intended to provide a city government with a 'systematic process for gathering information about the big picture and using it to establish a long-term direction and then translate that direction into specific goals, objectives, and actions' (Poister and Streib 2005, 46). Strategic planning emerged from the private sector management field and has been adopted by the public sector in many countries. It has been a key component of city governments adopting NPM approaches that are viewed as promoting improved performance. Strategic management and or strategic planning requires considerable time and resources to gather data and identify local issues, undertake community consultation, and develop relevant solutions that will lead to improvement in the way city governments are managed and deliver their services. The capacity to pursue a strategic approach may vary according to geographical and socio-economic context. For example, well-resourced metropolitan councils may not face resource constraints of their counterparts in smaller or rural locations (Gordon 2013). Surveys of local councils suggest that while many have established strategic plans, the utilisation of the benefits of these plans and the use of comprehensive strategic management has only developed in a small number of leading-edge municipalities (Poister and Streib 2005).

More relevant to climate policy is the finding that there is lukewarm support among city government managers for linking performance management to strategic goals (Ammons and Roenigk 2015). This finding has important implications for the mainstreaming of climate policy, as discussed in Chap. 3, to achieve the objectives of climate action plans. Without the appropriate linkages, there will be challenges in identifying policy successes and failures to provide opportunities for policy learning.

One possible explanation for city managers' reluctance to link performance management to strategic goals can be found in the common view that public sector goals can be ambiguous and too difficult to quantify and measure (Rainey and Bozeman 2000). In the public sector, the common perception is that public policy is characterised by goal ambiguity rather than by clarity. Observers argue that the public sector needs to deal with multiple, diverse, and vague objectives when compared with private sector firms (Pandey and Rainey 2006). Such a view fits with the character of climate policy as a wicked problem that is complex and requires multidisciplinary approaches. The evidence seems to be clear that goal ambiguity emerges in government agencies where more 'complex, controversial missions and goals face more challenges, which in turn produces more problems with red tape, communication, commitment, and motivation' (Pandey and Rainey 2006, 100). Also concerning is the observation that ambiguous goals can harm individual and organisational performance (Jung 2013). Despite these challenges, there seems to be some agreement among public sector managers that goal clarity is needed within public sector organisations for objectives to be achieved (Rainey and Bozeman 2000). Many senior managers recognise that employees need to understand how their work connects to an organisation's near- and long-term strategic interests and objectives. Clear goals can promote higher levels of motivation among policy practitioners by encouraging them to exert greater effort and energy towards goal attainment (Locke and Latham 2002). Moreover, well-specified goals and goals created jointly between managers and their staff can be useful in measuring and assessing employee performance. The public management research confirms that 'job-goal specificity and mission specificity have a positive influence on the work motivation of employees in public agencies' (Jung and Lee 2013, 792). Importantly, clear goals provide employees with a mark of what is expected of them in their jobs as well as how performance will be evaluated and rewarded (Davis and Stazyk 2015).

Within the public sector, alignment between politicians and senior managers in terms of coordination and cooperation is fundamental to the effectiveness of a performance management system. Senior managers must be the main point of contact between politicians responsible for identifying goals and strategies, and the bureaucracy that is responsible for achieving them (Walker et al. 2011). Research shows that a government agency with a lower level of political support tends to have more disagreement among important stakeholders over its mission and goals; hence, less political support is associated with more organisational goal ambiguity (Stazyk and Goerdel 2011). This same research also suggests that agencies with higher levels of political support are more likely to have clear goals. The performance-related benefits and costs of political support appear to be carried through, at least in part, the level of goal clarity present within an organisation (ibid.). Such findings suggest that, in a policy area such as climate change, the importance of political support for the activities of organisational units cannot be overstated. With this support, organisation units within a city government will be in a position to identify what tasks they are responsible for achieving, and which domains they need to operate across. Without political support, city government senior managers may be uncertain about what they are expected to achieve in terms of the climate objectives outlined in a strategic plan.

In the context of climate change policy, more work needs to be done to identify approaches by city governments that promote an improvement in the linkages between politician decision-makers and senior managers in establishing and implementing strategic goals to a performance system. In addition to political support, improved communication and managerial autonomy and authority can also contribute to an improvement in goal clarity. Research suggests that any effort to clarify organisational goals, combined with increasing the level of managerial authority within an agency, is likely to result in better organisational performance (Jakobsen and Mortensen 2016). Surveys of public sector managers show that the adequacy of communication along the various dimensions leads to organisational mission and goals that are clearer (Pandey and Garnett 2006; Stazyk and Goerdel 2011). We can conclude from the current research that when implementing climate policy, city governments need to make every reasonable effort to clarify their mission and goals through such procedures as strategic planning and investment in improving internal communication. Important questions to ask in this regard include whether the city government's climate change mission is clear for the organisation as a whole and for subunits. There must also be clarity in political climate change objectives and management's information and communication concerning these objectives in the context of the city government's strategic plan (Jung and Lee 2013).

The need for goal clarity also has relevance for city governments participating in networked arrangements with other city governments or with local stakeholders. When public organisations are participating in networks, research suggests that they probably have more goal-related demands placed on them, and those organisational and political processes utilised to merge and prioritise network and organisational goals may not always be successful (Davis and Stazyk 2015). The current research on the 'orchestration' of goals by city network alliances raises question about the degree to which individual members are being steered towards particular actions and objectives (Gordon and Johnson 2017). As a consequence, organisational goal ambiguity can increase both generally and for employees working with these network arrangements, where the goals of the network may not align with those of a specific city member. However, while engaging network partners may be costly for the organisation, it has the potential to pay the agency back in the form of increased performance (Davis and Stazyk 2015). To the extent that clearer goals and roles contribute to organisational performance, city government practitioners should strive to understand the mechanisms designed to maximise goal and role clarity. For city governments serious about climate change, the networked nature of policy implementation seems to be unavoidable, limiting the tensions between networks. Public managers' favourable interpretation of their organisation is a critical area for future research in public management.

Measurement

Measuring performance involves the systematic collection of data by observing and registering performance-related issues for a performance-related purpose (Bouckaert and Halligan 2008). The performance measurement process depicts measurement as consisting of five distinct and chronological steps: targeting, indicator selection, data collection, analysis, and reporting (Van Dooren et al. 2010, 55). One of the underlying assumptions of performance management is that performance data is based on relevant good-quality measures and should have 'purpose' so it can be used to better allocate resources, make decisions about strategy, re-engineer processes, motivate staff, and facilitate accountability (Moynihan, 2009; Kroll 2015). The importance of goal clarity is reinforced here, as measurement will be based on the clearly understood political objectives for climate policy. Goal displacement or the subversion of goals has been identified as a major limitation on performance measurement (Bianchi and Williams 2015). Clarity should help to reduce the risk of a common tendency to focus outputs at the expense of outcomes (Bohte and Meier 2000), thus limiting the potential impact of climate initiatives.

City governments regarded as exemplary in their performance differ from all cities on the character of the measures they use, whether they benchmark them, set targets for performance, and report on them frequently (Sanger 2013). The promise is that measurement, the quantitative representation of the quality or quantity of inputs, outputs, and/or outcomes, can offer valuable data for policy-makers by assisting with comparison, prediction, and evaluation (Sterck and Scheers 2006). This approach is of particular relevance to climate change mitigation policies for city governments' need to demonstrate linkages between their policies and reduced emissions (Sippel 2011; Kennedy et al. 2012; Garren and Brinkmann 2012). Providing accurate, timely, and relevant data in formats pertinent to the needs of stakeholders is a critical foundation for city government climate policy. Information deficiencies can contribute to political resistance and public backlash particularly where city-level policy impacts on lifestyles and behaviour change.

Priorities of measurement efforts will in large part be determined by the uses of the performance information. City managers need to identify what the performance information will be used for in terms of the implementation and evaluation of climate policy. The performance literature shows that measurement must produce information that contributes to improvement through better informed decisions, goal-based learning, and sanctioning and/or rewarding (Kroll 2015). Policy learning is the key objective, and measurement data must produce information that keeps the climate policy both on track of current activities and concerned with future improvements (Van Dooren et al. 2010). Deciding what to measure will be limited by access to resources; as measurement is costly, some measurement may be undertaken by the state or national government as part of national programmes. For this reason, many cities are confined to measuring their own activities rather than the more difficult-to-measure community-wide initiatives. Unlike national-scale climate action planning and policy development, which can rely on well-established methods (UNFCCC 2007), for city-scale planning, there is a limited understanding of which option could yield the most effective results (Erickson et al. 2013).

The areas of responsibility for city governments and their political context will contribute to determining what they can measure. Areas of responsibility have implications for the first step in the process where city governments establish *targets* for their climate policy to achieve. Setting achievable targets for activities under their control that extend and improve current levels of performance is a critical foundation for measuring improvement. In terms of GHG emissions, city-level targets need more than political statements and aspirational goals. Many urban-scale climate action plans have focused more on initiatives that provide incremental and visible reductions in GHG emissions, such as data on the distribution of efficient lighting, upgrading of municipal vehicle fleets, or tree planting, than on longer-term, transformative, and potentially higher-impact actions such as altering urban forms or developing major new transportation systems (Betsill and Bulkeley 2007; Keirstead and Schulz 2010). Whichever approach is taken will have implications for performance measurement, as urban-scale targets for climate action need to be adopted through an assessment of whether the policies chosen are capable of meeting them (Erickson et al. 2013).

As the second stage in the measurement process, the selection of 'good' *indicators* needs to be determined by criteria that should include relevance, timeliness, and feasibility. A challenge for policy-makers is the identification and agreement on indicators for climate change performance measures that are both relevant and usable. Regular auditing and evaluation of performance measures establishes a quality assurance process of continuous improvement for measurement systems (Van Dooren et al. 2010, 111). This is of particular relevance to climate change policies for city governments, as evidence shows that some are able to demonstrate linkages between their policies and reduced emissions, while others have failed to establish usable indicators or demonstrate their contribution in increasing or decreasing GHG emissions (Sippel 2011; Kennedy et al. 2012; Garren and Brinkmann 2012).

An important element of good indicators is to establish a baseline from which performance for policy effectiveness can be measured. A GHG inventory is a common first step for city governments, as it serves as an indicator of emission-intensive sectors, as well as provides verifiable metrics upon which to facilitate targeted climate policies (Ibrahim et al. 2012). Inventories for cities require a combination of indicators of both production- and consumption-based emissions' attribution. While some emissions physically occur within the city boundary (e.g., transport emissions, fossil fuel combustion for heating, etc.), some emissions that are released outside the boundary are a direct result of urban activities (e.g., electricity generation and waste decomposition). GHG emissions can be associated with key urban materials such as food, water, fuel, and concrete (Ramaswami et al. 2008). As more of these consumption-related emissions are incorporated into a city's inventory, the inventory begins to reflect good indicators of a city's carbon footprint. Researchers argue that to facilitate benchmarking and policy development, a common approach to indicators is needed that meets the needs of cities, urban researchers, climate financing, and national monitoring and reporting requirements (Hoornweg et al. 2011; Kennedy et al. 2012). There are several international sources where community GHG inventory results for cities have been collected. These include the Carbon Disclosure Project (CDP) (2010) report on C40 cities, the 'Carbonn' Cities Climate Registry 2011 annual report, and the urban websites of both the UNEP and the World Bank.

Data collection can be through internal or external sources—while there are advantages and disadvantages to both, for example, cost, risks of gaming, and application restrictions, the information provided is a vital third component of the measurement process. For effective performance management, city governments need to undertake both quantitative and qualitative examinations that allow a deeper understanding of the factors that influence the operationalisation and integration of climate change policies into the overall strategic planning and implementation systems (Sippel 2011; Boyko et al. 2012). In many jurisdictions, city governments may have limited resources for data collection and analysis. Studies of small- and mid-sized cities in both the United States and New Zealand show that lack of funding and limited resource capacity are the main barriers to implementing sustainability policies (Laurian and Crawford 2016). Limited resources and lack of information are typical constraints faced by city governments in tackling climate change (Measham et al. 2010). Scientific data associated with climate change, for example, measurement data relating to fluctuations in GHG emissions, may be too costly for city governments to collect and analyse. Alliances with external stakeholders and the formation of partnerships are becoming increasingly common and assist city governments in addressing many limitations of measurement and the implications for reporting (PlaNYC 2007; Lutsey and Sperling 2008).

Data *analysis and reporting* is the final step in the measurement process that turns data into useful and relevant information. Reports need to be designed for use by stakeholders, such as citizens, senior managers, politicians, and media. Identifying who will use the information and deciding the most relevant format for reports are major considerations in achieving the overall objectives of performance management (Van Dooren et al. 2010, 54–75). Ongoing consultation with the above-mentioned stakeholders will be an important element in establishing what should be measured and whether performance information is useful and relevant. Evaluation is a key element of policy analysis that identifies where things are working and where there is a need for change. Performance measurement is therefore a key component of the evaluation process as policy-makers assess the outputs and outcomes of their programmes. Performance measurement should be seen as a work in progress, with feedback from auditing and evaluation processes contributing to improvements to the overall system. There is an ongoing need for more accurate qualitative analysis of how leading city governments are able to face the significant challenges of establishing performance management systems based on measurement data that accurately reflects the outputs and policy outcomes (Jones 2015). Relevant information is critical for internal stakeholders to meet the requirements of 'incorporation' of performance information into the management processes and procedures within city governments.

Incorporation

For measurement data to be used, deliberate action needs to be taken within city government management processes to incorporate performance information into policy, financial, and contract cycles. From an analytical perspective, there is a hierarchy between the cycles. Policies set out the priorities, which then translate into programmes and budgets. Only then does the question of which agency will perform which task arise. The policy cycle should determine the financial cycle, which then should determine the contract cycle (Van Dooren et al. 2010, 81). Incorporation helps to strengthen a performance system by introducing performance and measurement into the discourse and culture of the organisation. One of the key concerns is how organisations try to 'institutionalise performance data and information into recurrent, sustainable and systemically embedded practices that enforce performance based practices' (Bouckaert and Halligan 2008, 217). Embedding performance into management processes stands as an indicator of a government's commitment to performance as a tool to support policy implementation and accountability. Deliberate action is needed to incorporate performance into the operations of city governments. If performance is embedded, then information can be used in effective ways to measure results and support change when and where necessary.

The foundation for incorporation starts with a strategic plan that provides the vision and a guide of the policy process and includes objectives, performance targets, resources, activities, outputs, and outcomes (Bryson et al. 2010). City governments' strategic plans need to clearly articulate the climate change objectives. As outlined in Chap. 3, these plans form the basis of coordinated policy action. The challenge is in linking climate change and incorporating performance to help in mainstreaming climate policy across organisational units and functions. Mainstreaming is the key to ensuring climate change and sustainability are a consideration in each relevant policy area and in implementation processes across city government functions. By embedding performance across these functions, through performance-based management frameworks, city governments help ensure climate objectives form part of the evaluation process across their operations. Monitoring arrangements such as the balanced scorecard

(BSC) or the International Organization for Standardisation (ISO) can be used to guide incorporation. ISO 9000 has been popular with city governments in the United Kingdom in providing a framework for incorporating performance measurement into administration and management (Douglas et al. 1999).

Evaluation and feedback on the implementation of the strategic plan help to establish a feedback loop and provide evidence for the development of the next strategic plan. As an illustration, a strategic plan can establish the objective of improving air quality in the city within its overall climate change plan—see Fig. 4.1. By adopting a strategic planning process, some consideration of the need to mainstream climate policy across city government functions will be undertaken. The incorporation of performance establishes a process of monitoring and evaluation that provides feedback to reassess initiatives and provide reports for considering changes to the next strategic plan and re-evaluating the original goals and the assumptions behind them. The evaluation stage can include benchmarking performance against other cities, perhaps from a coalition of network cities. Surveys of local businesses and citizens for their views can also help clarify the local impacts of the climate initiative. Throughout this process, performance is incorporated in the planning, monitoring, and evaluation of the city's policy action. The objective of improving air quality will involve a range of policy initiatives designed to reduce GHGs in the atmosphere, for example, reducing emissions from buildings within the city boundaries. Air quality metres provide the raw data

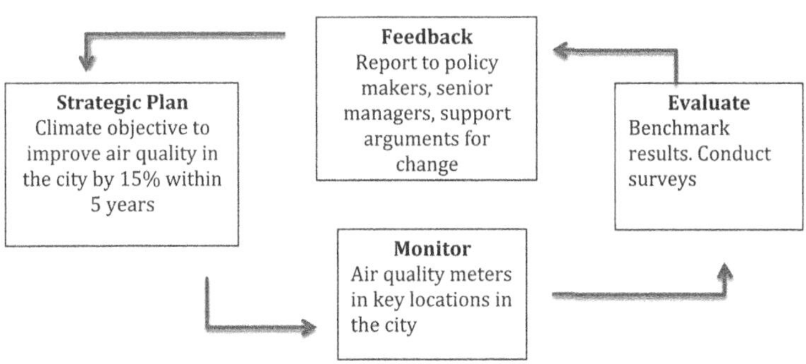

Fig. 4.1 Incorporating performance in the policy cycle

needed to build relevant information that can be evaluated in terms of the effectiveness of the approaches being taken.

Benchmarking facilitates comparison with other cities that may have chosen alternative approaches to gauge their potential in the local context. Surveys provide information that can evaluate the potential acceptance of alternative approaches within the local population. This material is collated into relevant feedback and provided to management committees from across the city government's organisations units in order to help identify alternative approaches that could be adopted by different functional areas. Incorporation helps build the information, which can then feed back into strategic planning processes and establish new approaches or strengthen existing targets and initiatives. Throughout this process, the performance information provides the measures for success and the drivers for considering changes to existing approaches.

The financial cycle is embedded in the policy cycle, with budget details authorising expenditures for implementation. The contract cycle is linked to the budget cycle and defines the agreements between the key actors involved in the delivery of government services. Budget allocations for climate initiatives will be incorporated into the financial cycle. Budgets should incorporate the same information from the strategic plan, authorising expenses during implementation (Van Dooren et al. 2010). A performance audit helps reinforce implementation by ensuring there is comparison between results and what is outlined in the strategic plan. The objective of such audits is to provide information on compliance, accuracy of financial projections, and the efficiency and effectiveness of the approaches taken in achieving the climate change plan. Contractual arrangements with staff and relevant internal and external stakeholders are also part of the policy cycle in terms of performance expectations for city government's outputs, and the contribution of the outputs to outcomes. Contracts identify expectations, define performance requirements, and provide the basis to monitor and evaluate arrangements to help ensure satisfactory performance against the objectives of the strategic plan. Reporting arrangements should monitor contract implementation, including the achievement of key performance indicators (KPIs) for both staff and stakeholders as outlined in the strategic plan. Reports on the financial and contractual cycles feed into the evaluation and feedback processes of the policy cycle.

An important component of the incorporation of performance is the devolution of authority to managers at departmental and programme levels. These managers need to be given the tools and discretion to pursue the city government goals in a 'manner they deem best and the freedom to innovate where necessary' (Ammons and Roenigk 2015, 519). When performance is effectively incorporated into management processes, executive oversight helps to ensure managers adhere to the overall climate objectives. With this oversight, executives can demonstrate their commitment to the climate plan and their interest in achieving its objectives. The incorporation of performance in management processes ensures executives play a major role and promoting a performance-based culture and reinforcing strategic goals and monitoring performance (ibid.). By incorporating performance information, the capacity of city governments to control their attainment of climate change goals is enhanced. In this sense, incorporation is almost literally a bridge between measurement and use (Van Dooren et al. 2010).

Use

The use of performance data is a critical issue for city governments seeking to measurably demonstrate their achievement of climate objectives. The effective 'use' of performance measurement data allows for internal and external comparison, which in turn supports policy learning and improvement. It is important to recognise the importance of 'use' in the overall context of performance management. The prime use of performance information should be to improve performance (Behn 2003). Drawing on this key purpose, leading researchers in the field argue that performance information must be used to promote learning, facilitate managerial direction and control, and ensure accountability (Bouckaert and Halligan 2008). In order to be most effective, performance information should be available across the organisation. Information should be used where necessary to upgrade systems to specific standards (single-loop learning) and to adjust standards (double-loop learning), or even for consistent adjustments to systems (Bouckaert and Halligan 2008, 28). The concepts of single-, double-, and triple-loop learning, established by

Argyris and Schon (1997), have an important link with performance information. Single-loop learning examines how a city government is attempting to achieve its climate objectives by asking questions about the effectiveness of the initiatives in the strategic plan. Double-loop learning will require municipalities to ask whether they have the appropriate climate goals; that is, do the results show they are capable of achieving the original goals or should they pursue alternatives? One of the benefits of double-loop learning is the opportunity for bottom-up communication and engagement. City governments can take the opportunity to seek feedback from citizens during the implementation stages of policy programmes so as to gauge effectiveness. The IPCC (2012) proposed that a triple-loop learning approach would be needed for climate adaptation issues that has relevance for city governments—see Fig. 4.2. For climate change issues, city governments' performance information provides the basis for double- and triple-loop learning to analyse lessons from failures and successes, and translate these into more informed decisions (Turner et al. 2016).

Despite the allure of these outcomes, city governments should not, however, be naïve about the redeeming qualities of performance information. Researchers also warn that managers can respond differently when confronted with performance measurement. Table 4.2 illustrates some of

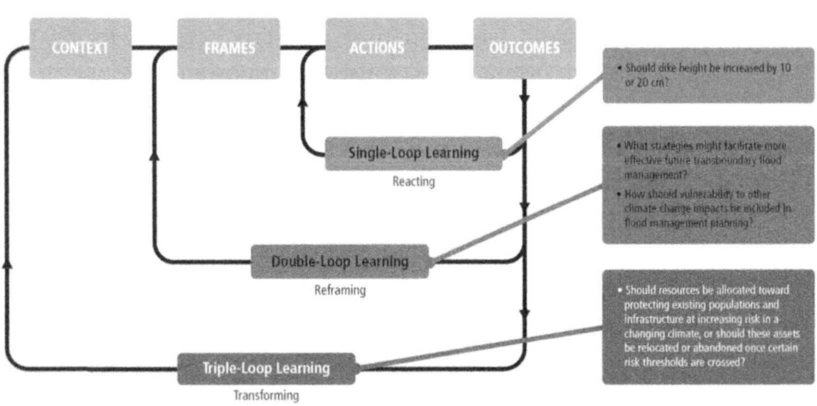

Fig. 4.2 Single-, double-, and triple-loop learning for climate adaptation
Source: IPCC (2012)

Table 4.2 Alternative uses of performance information

Managers might just use performance data passively (Moynihan, 2009), answering to stakeholders (Van Dooren et al. 2010), without drawing conclusions within their own zone of discretion (for an empirical study, see Moynihan and Lavertu 2012). Or they can use performance information politically (Moynihan 2009) to 'promote' (Behn 2003) their interest in turf wars, budget negotiations, or political arguments, where the interpretation of information is highly subjective, controversial, and role induced (for an empirical study, see Moynihan et al. 2011). Another type of use has been labelled dysfunctional or perverse (Moynihan 2009). It includes behaviours such as gaming, cheating, and cherry-picking (for empirical studies, see Bevan and Hood 2006; Kelman and Friedman 2009; Kroll 2015, 462).

the conclusions reached from assessments of the potential uses of information identified in public sector organisations. City governments need to be very clear about their intended use of performance information and take precautions to reduce the likelihood of such negative pathologies emerging.

The potential uses of performance information to meet the central purpose of improving performance should be factored into the management and measurement of climate change policy in city government operations and should consider the potential uses detailed in Fig. 4.3. A distillation of the extensive research literature on the uses of this information can provide much needed guidance for city governments. In essence, performance information should be used by city governments implementing climate policies in facilitating learning for internal and external stakeholders and giving an account of their actions. In terms of climate policy action agendas of city governments can move toward measurably demonstrating success by targeting measurement effort, providing guidance and direction through selecting performance indicators and transforming performance data into a form that supports effective management and reporting (Van Dooren et al. 2010, 100–112).

Reporting on progress of climate initiatives will be one of the most significant uses of performance information for city governments and their three key stakeholder groups: political leaders, managers, and citizens. Using information for reporting requirements can be imposed on city governments by political leaders or another level of government; in most cases, optimal indicators focus reporting on the outcomes of policy

1. strategic planning
2. setting programme priorities
3. adopting new programme approaches /changing strategies
4. clarifying objectives
5. performance budgeting
6. coordination of activities internally and externally
7. communication between managers
8. performance auditing
9. reporting and monitoring
10. evidence for compelling arguments for change
11. allocation of resources
12. evaluating outcomes of effectiveness
13. accountability to elected officials
14. formulating and monitoring contracted services
15. changing work processes
16. setting individual job expectations/staff performance plans

Fig. 4.3 Uses of performance information to incorporate performance in policy and management of climate policy

initiatives. External accountability demands high levels of validity and reliability in the reported information. These demand-side requirements support the learning objectives, which should feed back into the ongoing development and supply-side decisions on the development of the measures. Climate change reporting serves a number of internal purposes: First, reports are a key tool for politicians as policy-makers. Second, performance information can be part of compelling evidence to support the need for new policy action on climate change and/or to support actions already being taken to achieve policy outcomes.

The notion of evidence-informed policy has become a focus of attention, as policy analysts recognise that 'on any policy issue there is potentially a vast and diverse range of relevant evidence and many sources of informed advice' (Head 2013, 397). It is the role of policy bureaucrats within city governments to develop compelling arguments, with a combination of evidence from multiple sources and their own performance-based data, to reinforce or encourage change to the city's approach to climate change. An important observation is that politicians are pragmatic decision-makers, taking into account community sentiment and stakeholder values (Head 2013). If, however, policy-makers are determined to follow ideology and ignore the evidence, there is very little that practitioners can do to alter policy decisions (McCright and Dunlap 2003). Analysts refer to a 'policy and practice' pathway that 'helps policy actors navigate the uses of evidence' (Bowen and Zwi 2005, 166). Performance reports will be a key element in the evaluation of policy and contribute to the evidence of policy effectiveness. In term of climate change, policy evaluation should incorporate sustainability indicators to examine environmental, social, and economic factors, which in turn will contribute to more compelling arguments for political actors to consider (Herzi and Dovers 2006).

Reports also provide evidence for management and policy change within city government administration. As discussed in Chap. 3, an important audience for performance information reports are street-level bureaucrats within the city government. The implementation literature confirms the importance of these professionals, as their behaviour and judgement impact on policy outcomes. Performance information has a central purpose in motivating staff as well as in providing opportunities for them to learn and to improve their performance through examining different ways of undertaking their role (Behn 2003; Hupe and Hill 2007). Performance becomes a management tool when used for the purpose of motivation through performance budgeting and performance-based contracts or payment schemes. Performance information also provides critical information for understanding the implications of policy action or inaction.

Reporting is also critical for communicating with external stakeholders. In the past, governments have relied on mass communication strategies to raise awareness of the long-term consequences of unmitigated climate change. Research shows that these efforts have largely failed to encourage behaviour change in citizens (Moser 2006). Citizens need to be kept informed that the city government's climate policies are actually making a difference, that progress is being made towards the achievement of climate objectives. Benchmarking and other forms of comparison can be an effective analytical use of performance data for this purpose. Comparisons with past performance, other city governments, or with international standards can contribute to the effective use of performance information to communicate with citizens. Communication researchers have been recommending more tailored approaches that identify appropriate framings, messengers, and messages that are likely to resonate with citizens. Arguably, according to the subsidiarity principle, city governments, as the level of government closest to citizens, are in a better position than subnational and national governments to communicate more directly with local communities on climate change issues (Betsill and Bulkeley 2006). Different cities facing different climate change issues, for example, sea level rise or excessive smog levels, are being encouraged to tailor the use of performance information to be more effective in encouraging behaviour change. In New York, there are collaborations between the city government, the state government, and private metrics firms to produce reports regarding sea level and flood projections that will 'attempt to make it (adaptation) more tangible and inspire civic action' (Schiller 2016). Other examples show city governments establishing communication channels through engagement with citizens in forums for more direct dialogue and using communicators with persuasive power (Moser and Dilling 2011); the City of Sydney and the City of Greater Geraldton in Australia are two examples mentioned in Chap. 3. Other city governments, for example, Adelaide, Barcelona, Madrid, Rio de Janeiro, and Songdo in South Korea, are applying the use of performance information, in combination with technology, in a so-called smart city framework. Technology is being utilised to provide citizens with real-time data to encourage behaviour change in areas such as energy consumption, public transport use, water use, and waste management (Kitchin 2014).

Table 4.3 provides a summary of the key elements of the use of performance information by city governments relevant to climate change policy. The effective use of performance information can be the lynchpin in implementing climate policy, as it shapes what is measured and supports the effectiveness of the incorporation of performance information to promote policy and management change. From the literature, we can note that the central issues which city governments need to consider when attempting to accommodate the primary functions of using performance information are the questions about the intended use of the information, which instruments will maximise the effectiveness of performance to achieve the results being sought, the best data sources for measuring the

Table 4.3 Use of climate policy performance information by city governments

	To learn	Steer and control	Give account
Key questions	How to improve the management and implementation of climate change policy within organisational units	How to monitor and direct performance within the organisation in implementing climate policy	How to communicate results of climate change measures to internal and external stakeholders
Instruments	Strategic plan, climate change action plan, benchmarking with other city governments, business process re-engineering	Measurement scorecards, performance budgeting, performance pay	Public reports, performance contracts, citizen charters
Data sources	Technical measurement, surveys, statistical, international research institutions, self-assessments	Technical measurements, external observers, surveys, international and national data sets	Technical measurements, surveys, external observers, self-assessments, network databases
Report to whom	Staff, executive politicians, advisory boards, interest groups, citizens	City council, senior management	Citizens, executive politicians, subnational governments, funding bodies, international networks
Report format	Specialised reports, stakeholder persuasive testimonies, local forums	Scorecards, recurrent reports, testimonials	Annual reports, budget and accounts, specialised reports, interactive websites, persuasive testimonials, media releases, scorecards, local forums

Source: Adapted from Van Dooren et al. (2010; Chap. 7)

results intended, the stakeholders to which the reports should be directed, and the most appropriate format to communicate trends, key messages, and outcomes.

A Way to Assess Progress Towards Performance Management

Based on the material covered in this chapter, an analytical framework to examine city government climate action can now be proposed. The framework, outlined in Table 4.4, is based on the principles outlined in the performance doctrine and on the consideration of evidence from contemporary research on management practices for policy implementation. The performance use literature points to important drivers of data use that show significant positive effects when applied to the public sector. These variables are measurement system maturity, stakeholder involvement, leadership support, support capacity, innovative culture, and goal clarity (Kroll 2015). The framework is designed to allow the identification of observable evidence of these key drivers based on the four dimensions outlined in this chapter: first, *goal clarity* relating to climate policy; second, collecting and processing *measurement* data into relevant information; third, integrating *performance* into documents and embedding it into procedures and stakeholder discourses; and fourth, *using performance information* to improve decision-making, promote learning results, and support behaviour change and accountability. The framework will be applied through a case study analysis of city governments in the next chapters to produce insights into how performance systems are being applied through strategy documents, in management practices, and through the use of performance information. Reflecting on both the quality of the measures and their incorporation will facilitate an understanding of how performance information is 'used' for management and policy change towards the goals and objectives across city government departments and agencies, and relevant stakeholders.

Table 4.4 Observable evidence of performance management for climate change

Goal clarity

- The city government has a strategic plan that includes a response to climate change clearly outlining goals and objectives for all organisational units.
- The city has a climate change action plan that sets out clear objectives and targets in areas of local responsibility in responding to climate change.
- The climate change plan outlines actions to be taken, time frames, key performance indicators, and reporting requirements for each relevant organisational unit.

Measurement

- Data measures are developed and approved by internal and external stakeholders.
- The measurement system establishes qualitative and quantitative indicators for measuring mitigation and adaptation initiatives.
- Auditing and evaluation are carried out to monitor quality, usefulness, and relevance of the measurement information.
- Measurement is regarded as a 'work in progress' system to manage individual and organisational pathologies (e.g., gaming).
- Format and content of reports is appropriate for target audience.

Incorporation

- Performance is a mandated requirement for agency management of climate change initiatives.
- Performance management is implemented through a policy and/or strategic management framework.
- Performance is integrated into policy and management through policy, finance, and control cycles.
- Mechanisms to promote the matching of supply and demand of performance information.

Use of performance information

- Use of performance as a management tool, for example, monitoring unit and individual performance.
- A systematic use for comparing results, for example, benchmarking against other cities.
- Use of performance information for policy implementation and to learn how to improve policy and management.
- Efforts to communicate performance information externally.

A Staged Approach for Improvement in Performance

The performance management literature provides details on what good performance looks like in public organisations. The performance management doctrine shows that incentives and sanctions can contribute to improvements in the performance of individuals within the organisation and can serve as drivers of improvement. Increased transparency in the results of policy action contributes to strengthening accountability and citizen trust in the government. Policy learning resulting from the application of performance management principles reveals where improvements can and need to be made. However, these factors come into play when a decision at the highest level has been made that a performance-based approach will be adopted to improve policy success. City governments need to decide that a more effective approach is needed in responding to climate change, and that performance is the tool that drives the improvement.

The adoption of performance management as a tool for improvement can be detailed as a staged process where a combination of internal capabilities and systems and infrastructure can propel improvement. A staged approach developed by Mirvis and Googins (2006) provides a generative logic that drives change within organisations. This has some relevance to the management of city governments, as it is based on attitude change at the highest levels of decision-making in response to contextual challenges. In the case of city governments, the political and institutional context has been shown in Chap. 3 to be a determining factor in city governments' response to climate change. Political leadership has been shown to be instrumental in bringing about change—when leaders are prepared to champion a climate agenda, they increase the possibility of a policy window emerging (Dietz et al. 2007). Examples could include policy change by a new governor/mayor, demands from other levels of government, environmental crisis from extreme events, demands from the community. These factors will be context specific, so they will vary between cities and across time. A city may change direction often under new leadership, with new policies calling for new systems and infrastructure within the city administration, thus enabling a performance-based approach.

For Mirvis and Googins (2006), contextual 'triggers' will call for a fresh managerial response to environmental factors. For city governments, responding to these triggers can result in action that is progressively more effective and elaborate. Performance-based responses to climate change are an attractive option for city governments seeking to measurably demonstrate the effectiveness of their response. A staged heuristic is proposed here to assist city governments that seek guidance on areas where improvements need to be made to maximise the benefits of performance management to achieving success in climate policy. The contextual socio-economic, environmental, and institutional forces experienced by city governments will contribute to a response that varies from a minor involvement to comprehensive participation. The response can be assessed against the key elements of the performance doctrine of goal clarity, measurement, incorporation, and use of performance information. Progress through the stages means stronger adherence to the principles of performance management as a means to improve transparency and accountability for climate action.

Mirvis and Googins argue that 'movement along a single development path is not fixed nor is it attaining a penultimate "end state" a logical conclusion' (2006, 107). As this chapter has outlined, there is a broad scope of benefits from improving performance and reporting progress. With transparency through appropriate communication and reports, governments are more open to demonstrating real value to local residents, thus helping to build political and civic support for efforts to reduce the impacts of climate change on the city (Sanger 2008, s71; Whitmarsh et al. 2012). The stages show where the changes are needed to improve the application of the dimensions of performance as outlined in this chapter.

City governments may choose to respond to specific climate issues or specific factors contributing to climate change. For example, they may pursue specific policies in areas under their responsibility and control to mitigate GHG emissions. The strength of the response can also be determined by the urgency or significance of the issue; for example, a flooding event may prompt a focus on adaptation measures within a specific area of the city. A stronger, more performance-based approach will be possible in areas where the city has institutional control over legislation and has access to resources. The stronger the response, the more effective the performance-based response can be in contributing to improving transparency and accountability for the policy action taken by the city. Table 4.5 outlines a heuristic for the staged approach to the application of performance management by city governments responding to climate change.

Table 4.5 Proposed staged heuristic of performance management for climate policy

Stage	Goal clarity	Measurement	Incorporation	Use
1. Minor involvement	Aspirational, voluntary approach, vague objectives	No measurement commitment, no stakeholder consultation	No specific responsibility for performance	Minor reporting, vague in specifics
2. Limited commitment	Mix of mandated and voluntary targets in select responsibilities	Measures introduced in chosen policy areas with limited stakeholder consultation	Incorporation of performance in specific functional areas	Reporting limited to areas that show progress
3. Comprehensive participation	Mandated targets in areas of responsibility relevant to climate action	Specific measures in all areas established with extensive stakeholder consultation	Incorporation of performance across all targeted areas with relevant organisation units	Public reporting in all aspects of policy action designed for relevant stakeholders

In considering progress through the stages, city governments need to account for their history, politics, institutional arrangements, and civic infrastructure. Research shows that a performance management champion within the city government is essential to make progress (Moynihan 2008). The likelihood that resources, capacity, and expertise can be committed when transferring from one stage to another will be strongly influenced by the level of power of the champion. No two cities will be the same, and there will be potential for uneven progress across the stages, for example, despite having clear goals, a city may be reluctant to use performance information to report a lack of progress. For these reasons, there must be an acceptance within the city leadership that progress takes time, that progress remains uneven, but commitment from the highest levels, both political and management, needs to be long term (McDonald et al. 2003).

The next three chapters will present the case studies that highlight different dimensions of the application of performance measurement and management by city governments. By applying the analytical framework, we will gain greater insights into the contextual factors that contribute to the policy response by each city. The material gathered through the analysis of primary documents, secondary research materials, and in-depth interviews will help identify the extent of the commitment to performance as a driver of policy and management change. From the analysis, we will be able to place each city at a stage within the heuristic and deepen our understanding of the changes required to strengthen the application of performance as a driver of change within the city administration.

The three cities have been chosen for their application of performance as a means to promote management and policy change when developing and implementing climate policy. Of particular interest are the issues identified as most relevant to the application of performance management, including goal clarity, measurement systems, and incorporation and the use of performance data in the development and implementation of climate policy. *Tokyo municipal government* has applied a performance-driven system in its cap-and-trade scheme that is widely recognised as a highly successful policy in reducing GHG emissions from buildings in the city. *Stockholm city government* has applied an integrated management system that applies performance as a means to facilitate policy coordination in its climate action policy. Finally, *Copenhagen city government* is recognised for its comprehensive, strategic approach to climate policy, which also includes a performance framework that monitors and reports progress. Each of these city governments has been an active participant in global networks that encourage other city governments to take action and respond to climate change. Each has received international recognition as a global leader and as an example to others on effective ways to respond to climate change.

References

Ammons, D. (2015, December). Getting Real About Performance Management: Using Performance Information to Improve Services Is Key. *Public Management*. ICMA Publications, PM Magazine. http://icma.org/en/Article/106489/Getting_Real_About_Performance_Management

Ammons, D. N., & Roenigk, D. J. (2015). Performance Management in Local Government: Is Practice Influenced by Doctrine? *Public Performance & Management Review, 38*(3), 514–541.

Argyris, C., & Schön, D. A. (1997). Organizational Learning: A Theory of Action perspective. *Reis, 77/78*, 345–348.

Bains, M. A. (1972) *The New Local Authorities: Management and Structure*, The Bains Report. London: HMSO.

Behn, R. D. (2002). The Psychological Barriers to Performance Management: Or Why Isn't Everyone Jumping on the Performance-Management Bandwagon? *Public Performance & Management Review, 26*(1), 5–25.

Behn, R. D. (2003). Why Measure Performance? Different Purposes Require Different Measures. *Public Administration Review, 63*(5), 586–606.

Betsill, M. M., & Bulkeley, H. (2006). Cities and the Multilevel Governance of Global Climate Change. *Global Governance: A Review of Multilateralism and International Organizations, 12*(2), 141–159.

Betsill, M., & Bulkeley, H. (2007). Looking Back and Thinking Ahead: A Decade of Cities and Climate Change Research. *Local Environment, 12*(5), 447–456.

Bevan, G., & Hood, C. (2006). What's Measured Is What Matters: Targets and Gaming in the English Public Health Care System. *Public Administration, 84*(3), 517–538.

Bianchi, C., & Williams, D. W. (2015). Applying System Dynamics Modelling to Foster a Cause-and-Effect Perspective in Dealing with Behavioural

Distortions Associated With a City's Performance Measurement Programs. *Public Performance & Management Review, 38*(3), 395–425.
Bohte, J., & Meier, K. J. (2000). The Marble Cake: Introducing Federalism to the Government Growth Equation. *Publius: The Journal of Federalism, 30*(3), 35–46.
Bouckaert, G., & Halligan, J. (2008). *Managing Performance: International Comparisons*. Oxford: Routledge.
Bowen, S., & Zwi, A. B. (2005). Pathways to "Evidence-Informed" Policy and Practice: A Framework for Action. *PLoS Med, 2*(7).
Boyko, C. T., Gaterell, M. R., Barber, A. R., Brown, J., Bryson, J. R., Butler, D., Caputo, S., Caserio, M., Coles, R., Cooper, R., & Davies, G. (2012). Benchmarking Sustainability in Cities: The Role of Indicators and Future Scenarios. *Global Environmental Change, 22*(1), 245–254.
Bryson, J. M., Berry, F. S., & Yang, K. (2007). The State of Public Strategic Management Research: A Selective Literature Review and Set of Future Directions. *The American Review of Public Administration, 40*(5), 495–521.
Bulkeley, H., & Betsill, M. M. (2013). Revisiting the Urban Politics of Climate Change. *Environmental Politics, 22*(1), 136–154.
Christensen, T., & Lægreid, P. (2010). Increased Complexity in Public Organizations—The Challenges of Combining NPM and Post-NPM. In P. Laegreid & K. Verhoest (Eds.), *Governance of Public Sector Organizations: Proliferation, Autonomy and Performance* (pp. 255–275). Basingstoke: Palgrave Macmillan.
Davis, R. S., & Stazyk, E. C. (2015). Examining the Links Between Senior Managers' Engagement in Networked Environments and Goal and Role Ambiguity. *Journal of Public Administration Research and Theory*. DOI 10.1093/jopart/muv023.
Dietz, T., Dan, A., & Shwom, R. (2007). Support for Climate Change Policy: Social Psychological and Social Structural Influences. *Rural Sociology, 72*(2), 185–214. DOI 10.1526/003601107781170026.
Douglas, A., Kirk, D., Brennan, C., & Ingram, A. (1999). Maximizing the Benefits of ISO 9000 Implementation. *Total Quality Management, 10*(4–5), 507–513.
Erickson, P., Lazarus, M., Chandler, C., & Schultz, S. (2013). Technologies, Policies and Measures for GHG Abatement at the Urban Scale. *Greenhouse Gas Measurement and Management, 3*(1–2), 37–54.
Gahan, P., Adamovic, M., Bevitt, A., Harley, B., Healy, J., Olsen, J. E., & Theilacker, M. (2016). *Leadership at Work: Do Australian Leaders Have What*

It Takes? Melbourne: Centre for Workplace Leadership, University of Melbourne. workplaceleadership.com.au/sal

Garren, S., & Brinkmann, R. (2012). Linking Greenhouse Gas Emissions to Greenhouse Gas Reduction Policies at the Local Government Level in Florida. *Greenhouse Gas Measurement and Management, 2*(2–3), 146–161.

Gordon, G. L. (2013). *Strategic Planning for Local Government.* ICMA Publishing.

Gordon, J., & Johnson, C. (2017). The Orchestration of Global Urban Climate Governance: Conducting Power in the Post-Paris Climate Regime. *Environmental Politics, 26*(4), 694–714.

Head, B. W. (2013). Evidence-Based Policymaking–Speaking Truth to Power? *Australian Journal of Public Administration, 72*(4), 397–403.

Heinrich, C. J. (2002). Outcomes–Based Performance Management in the Public Sector: Implications for Government Accountability and Effectiveness. *Public Administration Review, 62*(6), 712–725.

Hezri, A. A., & Dovers, S. R. (2006). Sustainability Indicators, Policy and Governance: Issues for Ecological Economics. *Ecological Economics, 60*(1), 86–99.

Hoornweg, D., Sugar, L., & Trejos Gomez, C. L. (2011). Cities and Greenhouse Gas Emissions: Moving Forward. *Environment and Urbanization, 23*(1), 207–227.

Hupe, P., & Hill, M. (2007). Street-Level Bureaucracy and Public Accountability. *Public Administration, 85*(2), 279–299.

Hvidman, U., & Andersen, S. (2014). Impact of Performance Management in Public and Private Organizations. *Journal of Public Administration Research and Theory, 24*(1), 35–58.

Ibrahim, N., Sugar, L., Hoornweg, D., & Kennedy, C. (2012). Greenhouse Gas Emissions from Cities: Comparison of International Inventory Frameworks. *Local Environment, 17*(2), 223–241.

IPCC (Intergovernmental Panel on Climate Change). (2012). *Managing the Risks of Extreme Events and Disasters to Advance Climate Change Adaptation: Special Report of the Intergovernmental Panel on Climate Change.* https://www.researchgate.net/publication/254998982_Managing_the_risks_of_extreme_events_and_disasters_to_advance_climate_change_adaptation_Special_report_of_the_Intergovernmental_Panel_on_Climate_Change_IPCC

Jakobsen, M. L., & Mortensen, P. B. (2016). Rules and the Doctrine of Performance management. *Public Administration Review, 76*(2), 302–312. DOI 10.1111/puar.12409.

Jones, S. (2015). 'If You Can Make it Here, You Can Make it Anywhere': Performance Management and PlaNYC Climate Change Initiatives. *Regional Studies*. DOI 10.1080/00343404.2015.1052389.

Jung, C. S. (2013). Organizational Goal Ambiguity and Job Satisfaction in the Public Sector. *Journal of Public Administration Research and Theory*. DOI 10.1093/jopart/mut020.

Jung, C. S., & Lee, G. (2013). Goals, Strategic Planning, and Performance in Government Agencies. *Public Management Review, 15*(6), 787–815.

Keirstead, J., & Schulz, N. B. (2010). London and Beyond: Taking a Closer Look at urban Energy Policy. *Energy Policy, 38*(9), 4870–4879.

Kelman, S., & Friedman, J. N. (2009). Performance Improvement and Performance Dysfunction: An Empirical Examination of Distortionary Impacts of the Emergency Room Wait-Time Target in the English National Health Service. *Journal of Public Administration Research and Theory, 19*(4), 917–946.

Kennedy, C., Demoullin, S., & Mohareb, E. (2012). Cities Reducing Their Greenhouse Gas Emissions. *Energy policy, 49*, 774–777.

Kitchin, R. (2014). The Real-Time City? Big Data and Smart Urbanism. *GeoJournal, 79*(1), 1–14.

Kroll, A. (2015). Explaining the Use of Performance Information by Public Managers: A Planned-Behaviour Approach. *The American Review of Public Administration, 45*(2), 201–215.

Laurian, L., & Crawford, J. (2016). Sustainability in the USA and New Zealand: Explaining and Addressing the Implementation Gap in Local Government. *Journal of Environmental Planning and Management, 59*(12), 2124–2144.

Locke, E. A., & Latham, G. P. (2002). Building a Practically Useful Theory of Goal Setting and Task Motivation: A 35-Year Odyssey. *American Psychologist, 57*(9), 705–717.

Lutsey, N., & Sperling, D. (2008). America's Bottom-Up Climate Change Mitigation Policy. *Energy Policy, 36*(2), 673–685.

McCright, A. M., & Dunlap, R. E. (2003). Defeating Kyoto: The Conservative Movement's Impact on US Climate Change Policy. *Social Problems, 50*(3), 348–373.

McDonald, B., Rogers, P., & Kefford, B. (2003). Teaching People to Fish? Building the Evaluation Capability of Public Sector Organizations. *Evaluation, 9*(1), 9–29.

Measham, T. G., Preston, B. L., Brooke, C., Smith, T. F., Morrison, C., Withycombe, G., & Gorddard, R. (2010, June). Adapting to Climate Change

Through Local Municipal Planning: Barriers and Opportunities. *International Climate Change Adaptation Conference, Broadbeach, Australia, 29*.
Mirvis, P., & Googins, B. (2006). Stages of Corporate Citizenship. *California Management Review, 48*(2), 104–126.
Moser, S. C. (2006). Talk of the City: Engaging Urbanites on Climate Change. *Environmental Research Letters, 1*(1). DOI 10.1088/1748-9326/1/1/014006.
Moser, S. C., & Dilling, L. (2011). Communicating Climate Change: Closing the Science-Action Gap. In R. Norgaard, D. Scholsberg, & J. Dryzek (Eds.), *The Oxford Handbook of Climate Change and Society* (pp. 161–174). Oxford: Oxford University Press.
Moynihan, D. P. (2005). Why and How Do State Governments Adopt and Implement "Managing for Results" Reforms? *Journal of Public Administration Research and Theory, 15*(2), 219–243.
Moynihan, D. P. (2008). *The Dynamics of Performance Management: Constructing Information and Reform*. Washington, DC: Georgetown University Press.
Moynihan, D. P. (2009). Through a Glass, Darkly: Understanding the Effects of Performance Regimes. *Public Performance & Management Review, 32*(4), 592–603.
Moynihan, D. P., & Lavertu, S. (2012). Does Involvement in Performance Management Routines Encourage Performance Information Use? Evaluating GPRA and PART. *Public Administration Review, 72*(4), 592–602.
Moynihan, D. P., & Pandey, S. K. (2004). Testing How Management Matters in an Era of Government by Performance Management. *Journal of Public Administration Research and Theory, 15*(3), 421–439.
Moynihan, D. P., Pandey, S. K., & Wright, B. E. (2011). Setting the Table: How Transformational Leadership Fosters Performance Information use. *Journal of Public Administration Research and Theory, 22*(1), 143–164. DOI 10.1093/jopart/mur024.
OECD. (2009). *Measuring Government Activity*. Paris: OECD.
Osborne, D., & Gaebler, T. (1992). *Reinventing Government: How the Entrepreneurial Spirit Is Transforming Government*. Reading, MA: Adison Wesley Public Comp.
Pandey, S. K., & Garnett, J. L. (2006). Exploring Public Sector Communication Performance: Testing a Model and Drawing Implications. *Public Administration Review, 66*(1), 37–51.
Pandey, S. K., & Rainey, H. G. (2006). Public Managers' Perceptions of Organizational Goal Ambiguity: Analyzing Alternative Models. *International Public Management Journal, 9*(2), 85–112.

PLANYC (City of New York). (2007). PlaNYC: A Greener, Greater New York. New York City Mayor's Office of Long Term Planning and Sustainability, New York, NY.

Poister, T. H., & Streib, G. (2005). Elements of Strategic Planning and Management in Municipal Government: Status After Two Decades. *Public Administration Review, 65*(1), 45–56.

Pollitt, C., & Bouckaert, G. (2004). *Public Management Reform: A Comparative Analysis*. New York: Oxford University Press.

Rainey, H. G., & Bozeman, B. (2000). Comparing Public and Private Organizations: Empirical Research and the Power of the a Priori. *Journal of Public Administration Research and Theory, 10*(2), 447–470.

Ramaswami, A., Hillman, T., Janson, B., Reiner, M., & Thomas, G. (2008). A Demand-Centered, Hybrid Life-Cycle Methodology for City-Scale Greenhouse Gas Inventories. *Environmental Science and Technology, 42*(17), 6455–6461.

Sanger, M. B. (2008). From Measurement to Management: Breaking Through the Barriers to State and Local Performance. *Public Administration Review, 68*(s1), 570–585.

Sanger, M. B. (2013). Does Measuring Performance Lead to Better Performance? *Journal of Policy Analysis and Management, 32*(1), 185–203.

Schiller, B. (2016). Measuring What Exactly Will Be Destroyed as New York's Sea LEVELS Rise. https://www.fastcompany.com/3060604/measuring-what-exactly-will-be-destroyed-as-new-yorks-sea-levels-rise

Schultze, C. L. (1968). *The Politics and Economics of Public Spending*. Washington: Brookings Institute.

Sippel, M. (2011). Urban GHG Inventories, Target Setting and Mitigation Achievements: How German Cities Fail to Outperform Their Country. *Greenhouse Gas Measurement and Management, 1*(1), 55–63.

Stazyk, E. C., & Goerdel, H. T. (2011). The Benefits of Bureaucracy: Public Managers' Perceptions of Political Support, Goal Ambiguity, and Organizational Effectiveness. *Journal of Public Administration Research and Theory, 21*(4), 645–672.

Sterck, M., & Scheers, B. (2006). Trends in Performance Budgeting in Seven OECD Countries. *Public Performance & Management Review, 30*(1), 47–72.

Tucker, D. (1997). From Administration to Management. In B. Dollery & N. Marshall (Eds.), *Australian Local Government: Reform and Renewal* (pp. 69–88). Melbourne: Macmillan.

Turner, S., Fünfgeld, H., & Robertson, S. (2016). *Strategies for Embedding Climate Change Adaptation in Public Sector Organisations: A Review of the*

Academic and Grey Literature. Melbourne: RMIT University Centre for Urban Research.

UNFCCC. (2007). Synthesis and Assessment Report of the Greenhouse Gas Inventories Submitted in 2006. Note by Secretariat, FCCC/WEB/SAI/2006 28 June 2007.

Van Dooren, W., Bouckaert, G., & Halligan, J. (2010). *Performance Management in the Public Sector*. Hoboken: Routledge.

Walker, R. M., Avellaneda, C. N., & Berry, F. S. (2011). Exploring the Diffusion of Innovation Among High and Low Innovative Localities: A Test of the Berry and Berry Model. *Public Management Review, 13*(1), 95–125.

Whitmarsh, L., Lorenzoni, I., & O'Neill, S. (2012). *Engaging the Public with Climate Change: Behaviour Change and Communication*. London: Routledge.

5

City of Stockholm

Introduction

The City of Stockholm's 'integrated management system' (ILS) exhibits many of the fundamental elements of a performance management system. The ILS is a performance-based approach that establishes orientation goals and targets for the city government's operational areas, as well as performance indicators and activities to achieve policy objectives (City of Stockholm 2014). The city government argues that the ILS provides the means to eliminate conflicting objectives and ensure a sustainability perspective is included in all policy areas. In 2010, Stockholm became the first city to win the European Commission's European Green Capital Award, largely based on its 'long historical track record of integrated urban management' (EC 2010; Metzger and Olsson 2013, 2). The application process leading to the Award marked a turning point for the city government. The Award raised awareness at the political and administrative levels within the city of the work already being done in responding to climate change. This awareness created a policy window for local politicians and opened opportunities for new approaches to environment policies and more serious attempts to respond to climate change. As a result of its approach, Stockholm has taken a global role by being a

S. Jones, *Cities Responding to Climate Change*,
DOI 10.1007/978-3-319-64810-1_5

leading member of the C40 network and contributing as a 'pilot city' to the development of the *Global Protocol for Community-Scale GHG Inventories* (C40 2014). The C40 claims that Stockholm has surpassed many of its climate change goals, proving it has the 'political will and the technical experience necessary to achieve its ambitious targets of a renewable energy fueled future' (C40 2015, 75).

According to the City of Stockholm, per capita emissions in the city were reduced by 30% over the 2009–2007 period despite significant population growth and energy consumption increasing by 22% (City of Stockholm 2012a). Success in this regard has been achieved by two important factors: first, powering district heating and cooling through renewable energy, primarily biofuels (37%), fossil fuels (32%), and waste incineration (31%) (ibid.); and, second, low emission electricity supply, with 71% from wind, hydro, and nuclear sources (OECD 2013a, 22). The Mayor of Stockholm is a signatory to the European mayors' agreement, the 'Covenant of Mayors', which requires giving account annually to the EU of the actions taken and the results achieved in reducing GHG emissions in the city (City of Stockholm 2012a, 7). Stockholm scores well in the world's most liveable city indices and ranks in the top-five green cities in Europe in terms of energy and CO_2 emissions, transport, water, air quality, land use and buildings, waste, sanitation, and environmental governance (Economist 2012; Birch 2015).

Despite the international praises, there are concerns about external and internal factors that potentially limit the implementation of Stockholm's climate policies. Institutional issues that impact on the coordination between municipalities and the central government are seen to constrain both financial and regulatory options (OECD 2013a). Political tensions within the City of Stockholm have produced conflicts around energy and transport investments and climate policy (Rutherford 2014). Furthermore, the experience of senior managers reveals that the ILS, while being effective for vertical coordination within the Environment Department, has not produced the horizontal coordination necessary for mainstreaming climate policies across functional areas (personal communication, November 2016). Consideration of these external constraints and internal tensions, and how they are managed will be revealing in terms of implementing climate policy through a performance-based framework.

To better understand the details of the ILS and Stockholm's climate policies, this chapter will examine contextual issues. First, the anticipated impacts of climate change on the city are covered. The chapter then outlines the roles, responsibilities, and current trends in the local government in Sweden. More specific issues in terms of performance management will be covered in the third section by examining the approach by the City of Stockholm to strategic planning. This examination will provide insights into how the city has attempted to integrate climate change policies into the Council's overall responsibilities. Fourth, analysis of the nature of performance management within the ILS using the analytical framework developed in Chap. 4 will provide details on goal clarity, measurement, incorporation, and use of performance information in implementing climate policies. Finally, the chapter will draw some conclusions on the City of Stockholm's application of performance in the implementation of climate policy and ask critical questions about performance management in this European green capital.

Climate Change in Stockholm

Assessments by the Swedish government show that changes in temperature and precipitation across the country are in line with estimations in relation to global warming (Sweden 2014). According to the government, most activities in Sweden will be affected by rising temperatures and altered precipitation patterns. Of particular concern are the impacts on infrastructure for transport and communications. Buildings near lakes and watercourses will be exposed to flooding. Drinking water will be impacted by flooding, in addition to the potential for dam bursts, causing large-scale repercussions for energy from hydro systems (ibid.). Climate change will impact on Stockholm primarily through flooding from higher levels of rainfall and sea level rise and increases in temperature. Table 5.1 provides a summary of the likely impacts of climate change on the Stockholm region.

Table 5.1 Predicted climate change impacts in Stockholm County by 2100

Changes
• Average temperature increases of 4–6 °C
• Average precipitation increase of 10–30%
• Extended plant growth periods of 100–140 days
• Extreme weather events, for example, storms more common
• Number of snow days reduced by 65–100 days
• Flows in water bodies will increase greatly during winter, but reduce in summer
• Sea level rise up to one metre
Consequences
• Increased heat waves
• More favourable climate conditions for mosquitoes, tics, bacteria, and mould
• Increased flooding from sea level rise and rainfall
• Increased risk of landslides and erosion affecting built-up areas and infrastructure
• Risk of declining water quality
• Prolonged pollen season

Source: Adapted from: Colding (2013)

Flooding

The source of Stockholm's water supply, Lake Mälaren, will be subject to increased flooding due to heavier rainfall, and this will have a significant impact on buildings and industries in the city. Being a coastal city, of particular concern in Stockholm are the transport systems, where a one metre rise in the sea level will have considerable impact on infrastructure such as roads, railways, harbours, and airports, as well as water supply and power plants. In 2007, the Swedish government established the Swedish Commission on Climate Vulnerability to examine the country's threats and opportunities in facing climate change (Sweden 2007). The Commission raised the risk of flooding causing damage to the central functions of the network of underground tunnels (cable, car, and railway tunnels) or those adjacent to water (roads and railways) in the city. Stockholm's central train station and the associated subway network are particularly exposed. In a worst-case scenario, the Commission warned that flooding would impact railway traffic through the centre of Stockholm. Underground services would come to a complete halt, while bus services would not be able to run from existing city stations. A change in groundwater levels as a result of climatic changes could also pose a greater risk of landslides, erosion, and soil chemistry variations, with a potential impact on leaching of contaminations. In the event of predicted water level increases, along with extreme weather events, there will be impacts on the supply of electricity, telecommunications networks, and switching stations catering to central parts of Stockholm (Sweden 2007).

The predicted impacts of flooding caused by climate change have necessitated amendments to Stockholm's physical planning that give greater consideration to climate variations. In response to these calculations and predictions, the County Administrative Board recommended that no new constructions should take place in areas at risk of flooding within the next 100 years, where the likelihood of flooding is 63% or more. Hospitals, nursing homes, schools, or other significant infrastructure, water supply and sewage management plants, power plants, and telecommunications, as well as industries of a substantial environmental impact, should only be considered in areas that are not at risk of flooding (City of Stockholm 2007).

Temperature Increases

In Sweden, the densely populated areas, including Stockholm, have seen a shift from a cold-temperate to a warm-temperate climate, which reduces the frequency of winters bringing heavy snowfalls. Current projections suggest a warming in Sweden of greater magnitude and at a faster rate than global mean warming. A set of mid-range unmitigated climate change scenarios suggest a temperature increase of 3–6 °C towards the end of the century, depending on season and part of the country (OECD 2013b). The winter of 2007–2008 was the warmest of all winters since 1858–1859 in south-eastern Sweden. Calculations for Greater Stockholm estimate that, all other causal factors remaining stable, an increase of 4 °C to the mean summer temperature will increase mortality rates by 5%. Recent research has identified that heat extremes occurring in Stockholm in the 1980–2009 period increased the excess number of heat-related deaths due to climate change by as much as

55%, more than double the rate of mortality from heat extremes in this period than would have occurred without climate change (Åström et al. 2013). Temperature increase due to climate change is also expected to modify the distribution, seasonality, and incidence of infectious diseases (City of Stockholm 2007).

Roles, Responsibilities, and Trends in Local Government

The Constitution of Sweden establishes a two-tier system of local government (Chap. 1, Art. 7). As a reflection of its social democratic foundations, the constitution contains detailed guarantees of local self-government. There are 289 municipalities and 21 county councils in the region surrounding Stockholm city. The City of Stockholm municipality, with 900,000 citizens, is the largest municipality in Sweden. Municipalities are composed of popularly elected councils that collect income taxes and operate public services such as schools, child and aged care, utilities, housing, and cultural and leisure activities. The role of the Stockholm County Council is to supply the 26 municipalities, covering two million citizens in the Stockholm area, with relevant information, to coordinate central government interests, and to review local decisions in cases of complaint or on its own initiative. The County Council is also responsible for matters of common interest too extensive and too costly for individual municipalities to manage. This mainly concerns health care, public transport, support for business and industry, and regional growth and development. Regional economic development has largely been entrusted to state regional agencies. General supervision of land use planning and building rests with a national housing authority. The Stockholm County Council is the only council in Sweden to produce a regional plan. Figure 5.1 outlines the main objectives of the Plan and illustrates the range of strategies the County Council has established to achieve its goals.

International comparisons reveal that Sweden has, politically and functionally, the strongest local government forms in Europe (Wollmann 2004). Observers argue that Swedish local government does relatively

Fig. 5.1 RUFS 2010 Stockholm County Council Regional Plan
Source: Stockholm County Council (2013)

well in achieving policy coordination, democratic participation, and political accountability (ibid.). The autonomy of the municipalities rests on their power to levy their own taxes in the form of the local tax on earned income and pension income (Norton, 1994). Local government expenditures account for 55% of local taxes (Wollmann 2004). The overall volume of local government expenditure is agreed annually through consultations between the central and the local government. Although municipalities are free to set their own levels of taxation and service charges, the central government has negotiated tax ceiling for local authorities since the 1990s.

The traditional voter distribution in Stockholm between political parties makes it necessary to have at least three parties in coalition in the City Council in order to form a majority. This has manifested itself in the past 10 years as an ongoing tension facilitating a shift back and forth between the left (focused on a mix of strong public sector, public ownership, and public participation) and the right (focused on a strong role for the market, deregulation, privatisation, and decentralisation). This is combined with a constant green association (focused on public ownership, green taxes, car restrictions, and environmental protection), tying both major parties towards an environmental centre (Dastur 2005).

Governance in Stockholm City Government

The *Local Government Act (1991)* gives Sweden's municipal councils autonomy to decide on the organisational structure of their respective local authorities. The legislation allows the municipal councils to transfer and to contract out local government functions to 'a (local government-owned) corporation, a trading partnership, an incorporated association, a non-profit association, a foundation, or a private person' (Chap. 3, Sect. 16). Since the 1990s, Swedish municipalities have adopted many aspects of the NPM approach. Some municipalities have outsourced the provision of services by private (commercial) as well as non-profit (voluntary) organisations. The Swedish national government promoted this development by obliging the municipalities to search for ways to reduce costs and improve the delivery of local services. The idea of result orientation has been promoted through managerialist concepts and has been widely adopted by municipal administrations. The utilities sector, which in the past was largely handled by the municipalities, came under privatisation pressure due to the availability of international providers and international competition. The delivery and supply of public utilities (energy, water, sewage, waste treatment), a traditional field for Swedish municipalities, has been supported by the market liberalisation policies of the European Commission

In 1991, the Stockholm city government established a municipal company, *Stockholm Stadshus AB*, to conduct municipal operations through a number of limited liability companies established by the city. These companies supply a broad range of services, including housing development, school buildings, water supply, port facilities, and parking operations and tourist information (Stockholm Stadshus AB 2015). The City Council appoints the Board of Directors, who are responsible for governance, planning, and oversight of the subsidiaries to ensure they achieve the Council's targets. The city government is the major shareholder of Stadshus AB and receives an annual dividend—one million Euros in 2014 (City of Stockholm 2015a). With this level of control, the city has maintained its influence in areas that can impact on climate and sustainability issues. One example is in housing. The city has been able to push a sustainability agenda that supports improvements in water use and heating through energy efficiency requirements. In the 1990s, like most Swedish municipalities, Stockholm also began to sell some of its assets or parts of them to private investors and providers. Swedish budgetary issues over this period proved a further stimulus for municipalities to sell off their assets. As a result of these developments, the City of Stockholm withdrew from a number of traditional municipal activities, including energy supply.

Political responsibility for strategic planning rests with the Stockholm's City Executive Board (Board). While the Council consists of 101 elected members, the majority are part-time unpaid positions. In the interest of representation, the Board comprises 13 members from both the majority and opposition parties. The Board has overall responsibility for ensuring that the Council decisions are implemented, followed up, and evaluated. The city's financial administration and long-term development are also the responsibility of the Board. Meetings of the Board are not open to the general public. Implementation of the majority of the climate change initiatives has been the responsibility of the Environment and Health Administration (EHA) Department. The overall approach, however, is that various aspects of the response to climate change are shared across all the Council committees. Stockholm's Executive Office is responsible for control, follow-up, and development of the city's operations and for ensuring that all the political decisions, including responses to climate change, are implemented and reported to the City Council.

Strategic Planning

Stockholm's City Council approved the plan entitled *Vision 2040—A Stockholm for Everyone* in October 2015 as a broad set of objectives for the city. The assumption of *Vision 2040* is that 'environmental problems can be addressed using improved knowledge and more efficient technologies' (Gunnarsson-Östling et al. 2013, 66). According to the city government, technology needs to be utilised so that 'all Stockholm residents can live an eco-friendly life' (City of Stockholm 2015a, 13). Sustainability is framed in terms of its economic value—'environmentally driven technology' will make life easier, boost Stockholm's global profile, and support a transition to a 'circular economy' with a focus on recycling and green technology (ibid.). In order for this to occur, the city government will use *Vision 2040* to 'generate clarity about long term ambitions, with both internal and external stakeholders' (City of Stockholm 2015b). The Council committees, boards, and operations take more specific decisions based on the objectives of the Plan. The ILS provides the important link between policy and finance agendas, and encourages the review of directions and targets through four-monthly reports and operational statements.

The Council regards the city as a role model in sustainable living for other cities from social, ecological, financial, and democratic perspectives (City of Stockholm 2015b). *Vision 2040* is the overarching document that covers the city government's sustainability approach to new residential developments, public transport, eco-smart technologies, and environmental and energy use requirements. It places climate change in the context of growth:

> The population of Stockholm is growing so the city needs more housing and to ensure accessibility and expanded social services. At the same time, Stockholm must protect its living environment and reduce its climate impact. (City of Stockholm 2015b, 27)

Concerns have been raised about the application of the concept of sustainable development in Stockholm's planning policies. Criticism has focused on the multidimensional nature of sustainability as a concept and the lack of clarity the Plan provides for decision-makers if there is conflict between development and sustainability (Gunnarsson-Östling et al. 2013; Mörtberg et al. 2013; Mahzouni 2015). Questions about exactly what precedence protecting the living environment as a sustainability measure has over the needs of the growing city are difficult to determine from the details available in *Vision 2040*. Such circumstances reflect the common dilemma for local governments in the conflict between regulatory responsibilities and development objectives. This dilemma contributes to tensions between conservative and green political parties in the city government. As an example, the Green Party and the Social Democrats have been more ambitious in establishing initiative-based approaches to reduce emissions. The conservative parties have preferred a regulatory approach that focuses on reducing emissions through the responsibilities of local authorities in Sweden to manage environmental issues at city level. A lack of precision in the intent of political leaders impacts on goal clarity, which can in turn create challenges for decision-makers within operational units of the city government. Where this occurs, the City of Stockholm utilises other guiding policies to help clarify goals. These additional policies are more precise in details on what the city government wants to achieve in terms of its sustainable environment objectives, including climate change initiatives. The *Environment Programme 2012–2015* helps to make these policies more specific and aligns targets and objectives with city government decision-makers and functional units within the organisation. The specific policy documents are outlined in Table 5.2. The policy documents most relevant to strategic planning and performance for the implementation of climate policy initiatives are covered in greater detail in subsequent sections of this chapter.

Stockholm City Plan: A Walkable City 2010 *(Promenadstaden)*

The Council argues that the City Plan acts as a strategic navigation tool for how the city government will meet its goals and provide guidance regarding its intentions and objectives for sustainable development. The Plan also provides comprehensive policy guidance for the city government's operational units with functional responsibilities. The city's main strategy follows an economic competitive advantage approach to build on what the Council regards as the capital's economic advantages, which are linked to environmental improvements. The Plan reflects many of the objectives of RUFS 2010, and as such, it links the city to the wider region in terms of sustainability objectives. For example, a key objective of the Plan is to increase access to the transport system while, at the same time, minimising the environmental impact of the transport sector. Objectives include reducing transport by private cars, increasing the use of public transport, and promoting cycling and walking. Each of these areas is outlined in greater detail in the *Environment Programme* and the *Action Plan for Climate and Energy*, including targets, measures and implementation proposals, responsibilities, and reporting requirements. These will be outlined in greater detail next.

Table 5.2 Guiding policies of the Stockholm city government for climate change

Swedish Climate Change Strategy 2001	Development according to local condition is an important adjunct to the national policy instruments. State support for local climate investment programmes should have long-term results beyond the projects themselves, in the form of interest, organisation and co-operation.
Integrated Climate and Energy Policy	• 40% GHGs from non-ETS sectors by 2020 (compared with 1990 levels) • 50% of energy consumption from renewable energy by 2020 • 20% more efficient energy use (compared with 2008) • 10% renewable energy in the transportation sector • By 2020: a phase-out of fossil fuels in heating • By 2030: a vehicle fleet that is independent of fossil fuels • By 2050: zero net emissions of GHGs in Sweden.
Environment Code	The Environmental Code requires permits for major activities that are expected to be environmentally hazardous. Evaluating the impact on GHG emissions forms a part of the procedure for permit appraisal.
Action Plan for a Fossil-Fuel independent vehicle fleet	The key principle is a GHG emissions price for the transport sector to encourage fossil fuel–independent fuels, as non-fossil fuel–based vehicles are becoming cheaper than fossil fuel–based vehicles. This is to be achieved through various policy instruments, measures, initiatives, and incentives.
Fossil-Free Sweden 2045	By 2045 at the latest, Sweden will have no net emissions of GHGs, after which negative emissions will be attained. An emission pathway with intermediate targets forms the government's planning and follow-up of climate policy. Periodic reports to the Riksdag, annually and every four years, will be required. A climate policy council will be established to examine the implemented policy on an ongoing basis and to assess whether, as a whole, it is compatible with the national climate objectives.
Vision 2040	City of Stockholm's target for socially, financially, economically, and democratically sustainable development over the coming decades.
Stockholm City Plan: A Walkable City 2010—Vision for the City Centre	Provides guidance regarding the city's intentions and objectives for sustainable development. The plan also serves as policy guidance for the entire city's areas and functional responsibilities.

(*continued*)

Table 5.2 (continued)

Environment Programme 2012–2015	The function of the environmental programme is to break down the environmental goals of the city into smaller clear goals that are easy to follow-up for municipal operations.
Stockholm Action Plan for Climate and Energy 2012–2015	This long-term perspective is essential in keeping the global frontrunner position that Stockholm has on the environmental stage and is a precondition for achieving its long-term targets for the city.
Stockholm Waste Management Plan 2013–2016	The purpose is to have a strategy for how the City of Stockholm and other actors work in the area of waste—in planning new areas, reviewing building permits, and supervising environmentally hazardous activity, as well as in determining how residents, businesses, and other organisations should manage their waste.
Stockholm Water Programme 2006–2015	Where changes are made in land and water areas, these will be designed with future climate changes in mind.
Green IT Strategy for the City of Stockholm	Green IT is a collective name for the measures designed to reduce environmental impacts with the aid of information technology (IT). It involves both using IT to reduce environmental impact and reducing the energy consumption and environmental impact of the IT sector as a whole.
Urban Mobility Strategy for Stockholm 2030	Provides the guiding policies for priorities in large and small decisions pertaining to the city's roads and streets to promote a more efficient, safe, attractive, environmentally friendly, and healthy Stockholm in line with the vision and the City Plan.
Stockholm's Parking Plan	The city will in large city building projects plan for reduced car use and good conditions for environmentally appropriate transports. The city should demand bicycle parking in prime locations and in sufficient quantities.
Cycling Plan 2012	Travel by bicycle should be doubled at the city's checkpoints as a result of the City's efforts. The proportion of pedestrian and bicycle travels should be increased to at least 20 % in the suburbs and at least 50 % in the inner city.

Environment Programme 2012–2015

The City of Stockholm outlined its climate policy goals in the *Environmental Programme 2012–2015* (the *Programme*):

> For the City of Stockholm, the climate question is a priority and the City has adopted ambitious goals for becoming climate neutral. In the year 2050, Stockholm will be a city free of fossil fuels. (City of Stockholm 2012, 17)

The *Programme* encompasses six overarching registries that connect to both *Vision 2040* and the regional and national environmental quality objectives. The main focus areas of the *Programme* are environmentally efficient transport, non-toxic goods and buildings, sustainable use of land and water, environmentally efficient waste management, and a healthy indoor environment. According to the Council, the *Programme* is based on the 'challenges existing today, among them the fact that Stockholm is an attractive and growing city where the needs of nature and people complement each other in an environment characterised by function, qualities and biodiversity' (City of Stockholm 2012, 4). The *Programme* is a city-wide regulatory document which breaks down the municipal council's operations goals into more manageable interim targets. The City Council decides on indicators to measure the targets for each field of activity. The municipal committees and boards decide on indicators to measure the interim targets of the *Programme* in their action plans. The city government Executive Office has the overall responsibility for following up on the *Programme* and reporting annually to the Council on progress. The *Programme* identifies the municipal committees and boards that have a key role in implementing and/or following up measures. Each registry contains specific interim targets that describe how the municipal committees and/or boards are expected to contribute to the goals of their particular operations. The Vice Mayor is responsible for both Environment and Traffic with respect to the political oversight of the implementation of the *Programme*.

In addition to the regulatory requirements of the national planning legislation, Stockholm has undertaken a number of measures to reduce energy use in the city. Targets and measures in the *Programme* were developed in consultation with internal and external stakeholders. Senior managers in the Environment Department appointed consultants and worked in collaboration with in-house experts to draft policy proposals. Targets were established in areas where the city government had direct responsibility. There was strong political support for these objectives through the active involvement of Vice Mayor Ulla Hamilton, who acted as the advocate for the policy proposals with other elected representatives (personal communication, November 2016). The focus of the *Programme* would be on the production of GHG emissions within the boundaries of the city. Emissions from consumption activities would not be part of the strategy and continue to be a source of contention with community surveys by the city government in 2013 showing that 'greenhouse gas emissions and wasteful consumption are seen by Stockholmers as major environmental problems. The proportion who see unsustainable consumption as a problem is increasing year by year' (City of Stockholm 2013).

As part-owner of AB Fortnum Varme, the City Council is jointly responsible for energy production in providing district heating. Heating is increasingly provided from electricity generated from wind power, biofuels, and the incineration of waste in compliance with national and EU objectives. The Vartaverket power station, which supplies the city with power, cut its use of coal with the help of government incentives, but has reached a stage where further cuts will warrant significant investment in renewable technology. The Council has undertaken to cut energy use in its own operations through efficiency measures. It has also established a specific objective for energy use in new buildings. The Council argues that limited opportunities in the energy sector mean that its mitigation focus will need to be on ambitious energy efficiency measures for its own operations and in the reduction of emissions from traffic in the city (City of Stockholm 2012, 17).

The main focus of mitigation measures is traffic, which the Council recognises is one of the city's largest environmental issues (personal communication, November 2016). It is a complex area that involves policy change from national and regional stakeholders who have regulatory powers and policy interests that impact on the transport networks in the city. Traffic and transport are issues that require behaviour change by many local residents. Climate change commitments through traffic-related initiatives are strong indicators of commitments by governments to reduce emissions. In this regard, the City Council has attempted to reduce the number of motor vehicles on the roads and provided support for technical solutions that reduce emissions. The city's most significant initiative is the introduction of the congestion charge by the national government in 2006. The toll charge at key points of entry and exit from the city centre had an immediate and substantial effect of reducing traffic by approximately 22%, which has been consistent over time (Eliasson 2014). The traffic in the inner city has been reduced by up to 15%, and this has contributed to a decrease in CO_2 emissions across the Stockholm County by 2–3% (ibid.). The reduction is considerable when considered in the light of continued population growth in the city. Figure 5.2 shows, that despite the population increase of 30% since 1990, emissions from transport in the city have continued to decrease. Senior managers in the city government argue that the increase in tolls introduced in January 2016 will show that further reductions will be possible through this initiative (personal communication, November 2016).

The Council's other traffic reduction initiatives have been designed as complementary measures to the tax-based incentives introduced by the national government. The national initiatives are tax-based incentives to improve fuel efficiency and rebates for alternative fuel–based vehicles. To facilitate these initiatives in Stockholm, the city government's *Traffic and*

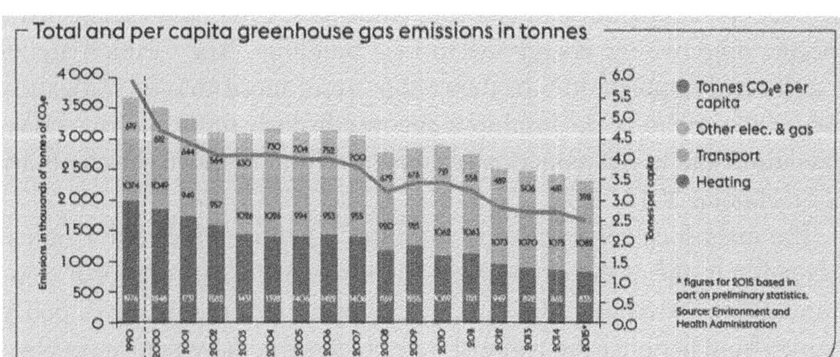

Fig. 5.2 Emissions reductions in Stockholm 1990–2015
Source: City of Stockholm (2016)

Waste Management Committee, the *City Development Committee*, the *Environment and Health Committee*, and the *City Planning Committee* have been responsible for implementing two key objectives relating to reducing emissions from the transport sector:

1. The City should act to make it easier to buy and charge electric cars. The efforts of the City have provided a foundation for developing an infrastructure for alternative fuels and a supply of green cars; and,
2. Sales of renewable fuels will increase and efforts to expand the number of tank stations providing alternative fuels will be prioritized (City of Stockholm 2012, 10).

In areas more directly under municipal control, the city government has set specific targets to improve the alternative forms of transport (cycling and walking) and to make public transport more attractive. The specific measurable targets for the 'Travel by foot and by bicycle' objective are the responsibility of the *Traffic and Waste Management Committee.* The *Programme* specifies two key targets:

> Travel by bicycle should be doubled at the city's checkpoints as a result of the City's efforts. The proportion of pedestrian and bicycle travels should be increased to at least 20% in the suburbs and at least 50% in the inner city. (City of Stockholm 2012, 9)

and:

> The proportion of public transport for mechanized journeys (car or public transport) will be at least 78% in the inner city and 45% in suburban areas during rush hour. In measurements carried out by SL (Stockholm Public Transport), public transport covers 76% of mechanized trips in the inner city and 43% of regional trips. (City of Stockholm 2012, 9)

The Council's *Urban Mobility Strategy* (2012) reiterates and expands upon the objectives outlined in both *Vision 2040* and the *Programme*. The strategy also mentions a Stockholm Agreement (*Stockholmsforhandlingen*), which is the result of a negotiation between the Swedish government, the City of Stockholm, the Stockholm County Council and the other municipalities in the county. The Agreement relates to SEK100 billion for infrastructure funding, inlcuding road and rail, until 2021.

Within the transport objectives of the *Programme*, the Council argues it wants to create a sustainable transport system. A smooth running transport system is seen as essential to Stockholm's competitiveness in the global economy. The Council recognizes that the current transport system has a series of negative environmental consequences, including CO_2 from emissions. The Council argues that it can influence the need for transport and the conditions for different modes of transport, for example, through active parking policies, financial incentives, and local traffic regulations, through its responsibilities for road maintenance and through information and campaigns (City of Stockholm City 2012, 7). But these are not expanded on within the *Programme*. Instead, there is reliance on changing the Council's own fleet of vehicles to fuel-efficient alternatives, encouraging staff to travel less, encouraging travel by bicycle, and making public transport more attractive.

In terms of adaptation measures, the City Council largely implements national regulations. In Sweden, climate adaptation is the responsibility of specialised government agencies, including the National Board of Housing, Building and Planning. The *Building and Planning Act 2010* requires climate impacts and adaptation to be considered in planning and permits for development. Municipal governments are required to adhere to the following considerations:

- The substantial management of land, water, energy, raw materials, and environmental conditions
- The conservation and efficient use of energy and water
- Chances of preventing water and air pollution

The *Programme* is integrated into the city's ILS system of governance and follow-up of operations. The integration means that execution and follow-up of the interim targets of the *Programme* take place in the action plan of the respective municipal committee or board. The *Programme* document does make it clear that all municipal committees and boards should take measures that lead to target fulfilment within their own operations, regardless of whether they have been given a specific interim target in the environmental programme. All municipal committees and boards are also bound by the environmental programme to clarify their own environmental impact and describe ways to reduce it in an environmental action plan. There is no guidance within the *Programme* for committees that face the environmental sustainability–development dilemma. Implementation could prove to be challenging for some, as many targets are vague statements of intent rather than regulated requirements of performance. For example, under the important objective dealing with the 'Proportion of environmentally certified buildings will be increased', the *City Planning Committee* has one of the weaker targets:

> A cooperative platform between municipal and other actors will be created with the goal of promoting environmental certification of buildings. (City of Stockholm 2012, 15)

The *Programme* is not clear on how committees and boards need to work with external stakeholders in achieving their targets. This seems to be left to individual discretion. The committees and boards with designated environmental responsibilities for an interim target are responsible for formulating an appropriate local target, as well as indicators and/or activities that follow-up the interim target of the *Programme*.

The Council also encourages wider exposure of committees towards achieving the objectives and targets. The Council suggests that the public should be able to follow the work of the committee in implementing the *Programme*. Since 2003, the 'environment barometer' on the city's website has shown progress against the targets, along with assessments and indicators (City of Stockholm 2017).

Stockholm Action Plan for Climate and Energy 2012–2015

The Action Plan for Climate and Energy 2012–2015 (the *Action Plan*) for climate and energy is closely linked to the *Programme* and the *City Plan*. The *Action Plan* is subordinate to the *Programme*. It defines more precisely the actions that can be decided by committees, or boards, and the City Council in order for targets of the *Programme* concerning energy and climate to be reached. The most significant target within the *Programme,* elaborated in the *Action Plan*, was the goal for Stockholm to be a 'fossil fuel free city by 2050' (City of Stockholm 2011). (In 2016, the City Government amended this goal to 2040: City of Stockholm 2016). The *Action Plan* identified 'conceivable' potential from expanding energy efficiency improvements beyond municipal property to all buildings in the city, and similarly for expanding the uptake of green vehicles beyond the city's own procurement policy (LSE 2013).

The climate and energy targets relate to mitigation initiatives in four sectors: eco-efficient transport, sustainable energy use in buildings, sustainable energy production in the Stockholm geographical area, and reduced energy use in city activities. The actions mentioned in the action plan aim to reach the goals of the *Programme* within these sectors. For targets where the City has authority of action, it is expressed as 'the City will…' Where the City lacks direct authority to act, the target is expressed as 'The City will strive to…' The action plan also suggests a number of knowledge-enhancing measures that will prepare the Council for upcoming action plans as well as for new climate and energy goals.

The most significant emissions reductions to 2015 were expected to come from continued improvements and expansion of district heating, and within the energy sector as a result of replacing coal with renewable fuel at the city's Värtaverket combined heat and power (CHP) plant (City of Stockholm 2011). Figure 5.2 shows the reduction in emissions and reveals that emissions from district heating reduced at a greater rate than other sources (City of Stockholm 2016). Other programmes in the transport and building sectors have contributed far less to emissions reductions in the short term. The largest reductions during 1990–2010 have been achieved in conversions from oil heating to district heating and a switch to biofuels in district heating (approximately 500,000 tonnes), conversions from oil heating to heat pumps (approximately 300,000 tonnes), replacement of fossil fuel–powered vehicles with green cars (approximately 80,000 tonnes), and replacement of diesel buses with ones powered by renewable energy sources in public transport (approximately 10,000 tonnes) (Stockholm 2012). The Council argues that after 2015, Fortum would not be able to achieve any major reductions of

GHGs from its energy production. As long as the coal-powered plant (KVV6) in Värtan is operational, GHGs of 260,000 tonnes will remain. It is considered technically very difficult to convert more than half of the coal to biofuels in this cogeneration plant.

Some of the proposed measures to reduce emissions have been designed to improve infrastructure for cyclists and pedestrians, improve public transport as an attractive alternative, support major railway investments, encourage alternative fuel vehicles, including both cars and trucks, and raise parking fees in the city. More details on the proposed measure can be seen in Table 5.3.

The *Action Plan* provides a series of ideas for Council committees to consider. It does not provide costings or time frames or measurement methodologies in association with each proposed initiative. The *Action Plan* is seen as an important basis for budget and operations planning for the city (City of Stockholm 2012, 7). The proposals are the result of consultations with the internal management group (personal communication 2016). External stakeholders consulted in the drafting process were primarily from national government agencies such as the Swedish Energy Agency and the Swedish Environmental Protection Agency. There is no indication of public consultation generally or with specific groups with an interest in sustainability. The City of Stockholm Annual Report for 2012 refers to citizen surveys showing that 82% of Stockholmers think it is good that the Council conducts active environmental work, and 76% feel that there should be tougher demands on citizens to live an eco-friendly life (City of Stockholm 2012a, 44). Annual reports in 2013 and 2014 provide partial updates of the progress made against many of the sustainability objectives of the *Programme*. There are ongoing improvements in the number of residents choosing cycling and walking, and the speed of public transport continues to improve. It is revealing that in 2016, senior managers in the Council agreed that changes to Council operations have proven to be more challenging than expected in areas such as green vehicles, energy and electricity consumption, and environmentally classed buildings (personal communication, November 2016).

A report conducted by the London School of Economics (2013, 8) claims that the City of Stockholm has achieved 'substantial success' in reducing emissions. However, the city would not reach its 2050 target

Table 5.3 Action Plan initiatives for reducing emission: heating, cycling, and public transport

Long-term goal	Proposed measures by City of Stockholm
Emissions from district heating be reduced by 50% The property stock of Stockholm be made 5% more energy efficient	The City strives, through its representation on the board of Fortum Värme, to ensure that the plans for conversion and extension are carried out according to the agreed schedule. Above all, focus is on the proportion of biofuels increasing as far as it is technically possible at the cogeneration plant (KVV6) at Värtan.
Increase the proportion of cyclists and pedestrians	• Conduct information campaigns about alternatives to cars • Provide improved information about bicycle paths • Expand system of lending bicycles • Improve maintenance of bike paths • Improve bike infrastructure with separate pedestrian lanes • Separate bikes from car traffic • Prioritise bike signals at intersections • Integrate bike paths in new development areas • Strive to combine bikes and public transport • Improve access to safe bike parking at traffic hubs
Make public transport more attractive	• Provide information and initiate projects for increasing use of public transport • Provide bus lanes for faster service • Use signalling priority in traffic control systems • Improve possibilities of switching between different types of traffic • Work for expansion of rail traffic • Implement Stockholm's Tramline and Trunk Network Strategy • Support the transition of the bus fleet to renewable energy • Investigate the possibility of boat commuting.

without 'strong and early policy action over the next few years to overcome long-term lock-in of high carbon infrastructure, systems and technology' (LSE 2013, 8). The report identified energy for heating and energy for transport and electricity supply as the key areas where reforms were necessary. While the City Council has considerable control in heating and transport policy areas, it would still require cooperation and coordination with the County Council and the national government to achieve the original target. In 2013, the OECD also examined Stockholm's record and recognised issues relating to the coordination between key stakeholders, particularly the central government, in streamlining policies and programmes (OECD 2013a, 10). Specific problems include the need for stronger public transport linkages, the need for greater discretion by the City of Stockholm in spending revenues from the congestion charge, and regulatory changes to facilitate renewable energy generators. These observations suggest a degree of frustration at the local level to institutional constraints within the political context that may limit the opportunities for the city government. At the time of writing this chapter, the city government released a new strategy that, it argued, would lead to Stockholm being fossil fuel free by 2040 (City of Stockholm 2016). In recognition of the issues associated with improved horizontal coordination of climate goals across the organisation and improved coordination with external stakeholders, the Council announced:

> A climate-smart Stockholm means incorporating the climate goals into all municipal operations and assigning a clear responsibility to certain municipal committees and the boards of certain municipal companies to coordinate and promote action. (Ibid., 7)

Similarly, the Council recognised the need to work closely with external stakeholders and improving its communication with local residents to promote behaviour change:

> The City needs to work with external players to achieve its fossil-fuel free goal. The Stockholm Climate Pact is a good example of long-term cooperation with trade and industry. In addition, residents are being informed through communication initiatives such as the 'Climate-smart Stockholmers' project that provides tools to help residents reduce their climate impact. (Ibid., 46)

It remains to be seen if future initiatives can improve the implementation of the climate change goals and achieve the long-term goal of a fossil fuel–free city by 2040. At this stage however, there are lessons for other city governments from the experiences of Stockholm in the context of the various aspects of the performance management framework developed in Chap. 4. The next section outlines this experience, with data gathered from interviews with senior managers from the city government staff in November 2016.

Performance Management and the City of Stockholm

Lessons from cities, such as Stockholm, well advanced in their climate agenda provide opportunities for reflection by other city governments contemplating climate policy initiatives. Experiences in policy development, implementation, and governance processes by leading city governments can show late adopters some of the challenges that can occur throughout the policy development process. In Stockholm, responsibility for the climate strategy action plan rests with the Environment and Health Administration (EHA).

Goal Clarity

The simply stated objective of a fossil fuel–free city by 2050 has been a clear goal for the city government since 2010. This goal has been supported through political commitment from both conservative and green political parties, which make up the coalitions governing the City Council. The consultation process in establishing the climate change action plan has helped to ensure internal stakeholder across the city government administration were well aware of the objectives that make up the climate strategy. The overall approach was to focus on developing targets that were achievable within the range of responsibilities and powers of the city government. As one senior planner commented:

> We were determined to focus on areas under our control and influence. We were careful to examine relevant issues through our consultation process so we didn't raise expectations that we were going to try to change the world overnight. (Personal communication, November 2016)

Consultations with external stakeholders, such as the Council agencies and private sector representatives, were required before the climate objectives were submitted to the Council for endorsement.

The fact that the Swedish government has a goal of being fossil fuel free by 2045 indicates broader political support for the ambitious goals of the city government and further raises awareness of climate goals in Sweden. The regular community opinion surveys issued by the city government also reveal a high level of awareness and support for the climate objectives. These surveys show that 80% of local residents support strong action by the city on environment issues, 70% support the congestion charge, 86% support initiatives to improve energy efficiency, and 58% support the waste recycling activities undertaken (City of Stockholm 2013). The 2014 election resulted in the 2016 review of the action plan, which has re-emphasised internal and external stakeholder support, with commitment to reduce the time frame for achieving the 'fossil fuel free by 2040' goal.

Senior managers within the EHA argue that the process for developing the action plan helped to ensure a high level of awareness and commitment to the climate goals. Each organisational unit was required to contribute to the development of the *Programme* by suggesting how their unit could participate in achieving the climate objectives over the life of the plan. Planners argued that the sign-off process raised expectations within the EHA that this would equate with agreement to achieve the targets set for each administration:

> In the early stages we were optimistic that when a manager signed off on the relevant section of the climate action plan that meant they would make the necessary arrangements to meet the objectives. (Personal communication, November 2016)

Sign-off from each city organisational unit was required before the final *Programme* was submitted to the Council for endorsement. The reporting process was also meant to follow the ILS and maintain focus on the goals. Environment Department officers concede that the process failed to produce the necessary coordination and achievement of the goals (personal communication, November 2016). Examples of the existing reporting format relating to improvements to public transport are highlighted in Fig. 5.3. One of the contributing factors was the aggregated presentation of performance data in the annual report, which allowed some of the poor-performing units to avoid exposure and accountability in contributing to the targets. The 2016 fossil fuel–free strategy outlines a range of solutions to the challenges the city faced when implementing the previous strategy (City of Stockholm 2016). Senior managers claim that new reporting formats and attempts to improve coordination across the organisation will improve mainstreaming by establishing more itemised and accurate reporting against climate and environmental objectives.

The goal of promoting sustainable housing has been one area of concern for senior managers throughout the implementation of the *Programme*. Sustainable housing has been an important element in supporting energy efficiency and reducing the environmental impact of the construction industry. The Hammersby Sustainable City development project was intended to be a model for sustainable development part of Sweden's bid for the Olympics in 2008. For various reasons however, Hammersby has been described as a failure (Iverot and Brandt 2011; Svane et al. 2011). Senior managers within the city government describe challenges of horizontal coordination in city administration, where professional and political agendas overrode the environmental objectives of the project. The outcome has been a poor result for the city, as described by Iverot and Brandt (2011), where 'a clear structure of the assessment process in the environmental programme, which would ensure the quality of gathered data and facilitate the development of even better sustainable urban districts in the future' (ibid., 1043). Importantly, further research shows that the coordinated project organisation did not strive to improve the formulation of the *Programme* in order to make it more authoritative as the project proceeded (ibid., 1055). Despite these failings, there is some evidence of policy learning. An attempt by the Environment Department to overcome the weaknesses of Hammersby includes a more project-focused approach, with early coordination across the city administration. Senior managers meet every two weeks to clarify issues and solutions through improved coordination of expertise in the development of the new Royal Seaport sustainable city (personal communication, November 2016).

Measure	Timetable	Prognosis	Year	Reduction of greenhouse gases (co$_2$e)	Energy efficiency
Strive for increased accessibility, partly by increasing the proportion of bus lanes	Investigation of more driving lanes for public transport 2010. Bus lanes for a large part of the trunk lines 2015.	Based on traffic mileage without measures	2015 2020 2030	1,840 tonnes 3,870 tonnes 4,180 tonnes	9.7 GWh 20.5 GWh 22.6 GWh
		Based on changes in the vehicle fleet	2015 2020 2030	1,530 tonnes 2,040 tonnes 1,670 tonnes	9.2 GWh 16.3 GWh 13.7 GWh
Support Stockholm Transport's planned change of the bus fleet	50% renewable 2011, 75% renewable 2016, 90% renewable 2020 and a fossil-fuel free fleet in 2025.	Based on changes in the vehicle fleet	2015 2020 2030	8,000 tonnes 23,000 tonnes 33,000 tonnes	13 GWh 51 GWh 111 GWh
Strive to realise the Tramline and Trunk Network Strategy	Investments according the the proposal are expected to be made in 2021–2030	Based on a new vehicle fleet	2021–2030	1,129–3,650 tonnes per year	6.6–13.9 GWh per per year

Assumptions for the calculations on increased accessibility:

- Average consumption for inner-city buses is equal to the consumption of buses in heavy city traffic.
- If the average speed increases from 12 to 18 km/h and the number of stops decreases from 6 to 3, fuel consumption can be reduced by 16 per cent.
- A reduced fuel consumption of 8 per cent is assumed for 50 per cent of the bus stock after 2015.
- A reduced fuel consumption of 16 per cent is assumed for 50 per cent of the bus stock after 2020.
- 50 per cent of the total driving distance for buses takes place in the inner city; it is primarily these that are affected by the measure.

Assumptions for the calculations on changes in the bus fleet:

- Change of the vehicle fleet is based on the goals of Stockholm Transport.
- The calculations start out from the prognosis of changes to the vehicle fleet and from Stockholm Transport's calculation that traffic mileage is reduced by 100,000 vehicle kilometres per 24-hour weekday if the City's and Stockholm Transport's Tramline and Trunk Network Strategy is realised.

Responsible	The City of Stockholm together with Stockholm Transport. Stockholm Transport places demands on subcontractors.

Measure	Timetable	Reduction of greenhouse gases (co$_2$e)	Energy efficiency
Example: A person cycles to work instead of driving, 7 km one way	1 year	3 kg per day 220 kg per summer (15 weeks) 672 kg per year (45 weeks)	10 kWh per day 770 kWh per summer (15 weeks) 2,300 kWh per year (45 weeks)
Example: 20 per cent of the increased number of cyclists by 2015 earlier drove	2012–2015	840–1,120 tonnes per year	3–4 GWh per year

Assumptions for the estimates

- Roughly 54,000 cyclists pass the inner city of Stockholm each day.
- The number of cyclists are assumed to increase by 10 per cent annually. 20 per cent of the increase is assumed to be transfers from car.
- The average journey distance is assumed to be 7 kilometres one way, according to the report »Work commuting in metropolitan areas – a situation analysis 2011:3« from Transport Analysis.

Timetable	Work in progress, should be intensified
Financing	Improved infrastructure, purchasing bicycle racks, land for bicycle parkings, etc. • Roughly 50 million kronor per year for new bicycle paths • Roughly 6 million kronor per year for maintenance of bicycle paths • Roughly 200,000 kronor per year for new bicycle parkings
Responsible	The City of Stockholm can strive to make sure the measures are implemented, and carry out information efforts

Fig. 5.3 Measures to improve attractiveness of public transport, including expected reductions in emissions, reporting from 2012 to 2014
Source: City of Stockholm (2011)

2012

Indicator	Goal	Outcome	Goal attainment	Comment
Percentage of cyclists	10%		No measurement taken	Value taken from travel habit survey in spring 2013
Metres of new cycle path in the city	Rise	5,230	Met in full	The outcome for the Development Committee is higher than the target value while the outcome for the Traffic and Waste Collection Committee does not meet the target value. Together they meet the City Council's aggregate goal of 5,200 m.
Number of cyclists	57,000	56,308	Met in full	The indicator measures an average over 5 years which gives an outcome below the target value. The value for 2012 only was 63,620.
Average speed of express buses in city centre	Rise	18 km/h	Met in full	Outcome in 2011: 17 km/h
Average speed of express buses in the outer districts	Rise	30 km/h	Met in full	Outcome in 2011: 29 km/h
Lengthening of travel time in %, city centre streets	Unchanged	-19%	Met in full	
Lengthening of travel time in %, inner approaches	Unchanged	-17%	Met in full	
Lengthening of travel time in %, connecting routes	Unchanged	-4%	Met in full	
Lengthening of travel time in %, outer approaches	Unchanged	-4%	Met in full	

2014

Indicator	Goal	Outcome	Goal attainment	Comments
Number of bicycle passages across the inner city	65,500	57,320	Met in part	This outcome is 88% of the annual target and is thereby assessed to be met in part. This indicator refers to the average of the past five years. The 2014 measurement took place in bad weather, which affected the outcome negatively.
Percentage of bicycle users	10.0%	10.0%	Met in full	This proportion is assessed to be unchanged compared with 2013.
Percentage of pedestrians	20.0%	20.0%	Met in full	-"-
Percentage of car travellers	24.0%	24.0%	Met in full	-"-
Percentage of public transport passengers	47.0%	47.0%	Met in full	-"-
Average speed of mainline buses in the inner city	18 km/h	16.8 km/h	Met in part	This outcome is a small increase over 2013 (16.5 km/h).
Average speed of mainline buses in the suburbs	31 km/h	28.4 km/h	Met in part	This outcome is a decrease compared with 2013 (30 km/h).
Increase in travel time, inner city streets, percent	0.0%	-5.0%	Met in full	Additional improvement after two years of lower travel times.
Increase in travel time, suburban streets, percent	0.0%	17.0%	Not met	After two relatively good years, the delays were again higher in the suburbs. This is normally due to more congestion, but the traffic measurements from October for the suburbs have not yet been compiled.
Travel time reliability	63.0%	63.0%	Met in full	

Fig. 5.3 (continued)

Housing is a significant issue in Stockholm, and the city government is a major provider of rental accommodation. The city maintains an active role in housing development. Providing housing is an economic issue, as workers attracted to the city complain of a lack of affordable housing, as they must compete with locals and migrants for housing. In 2016, the problem was exacerbated by the refugee crisis, as Sweden had accepted more refugees that most EU countries. The city government's leading role in housing development in the city means that it will be pursuing a sustainability approach to building, as this will help reduce their costs in the long term. A similar story is reflected in the cycling objectives. Despite strong political support and a significant allocation of funds, the conflict between competing goals of traffic planners and urban planners has seen a stronger focus on traffic management and retrofitting cycling infrastructure. The ILS was ineffective in overcoming the lack of horizontal coordination and goal conflict between these functional areas.

Measurement

The city chose the UN GHG Protocol as the basis for its measurement of GHGs across the city. The city has also been an active member of the ICLEI and the C40 by contributing to the development of the *Global Protocol for Community-Scale Greenhouse Gas Emission Inventories*. Members of the city government see participation in international measurement activities as important. The networks allow them to keep up to date with developments regarding the measurement of emissions and to contribute to the political agenda of raising the profile of city government climate change policy. Participation also produces co-benefits for the city. As one senior planner described the current situation:

> We are focused on benchmarking and a good response to reducing emissions against the performance of other cities can make us more attractive to investment. We are in a very competitive world for investment that helps drive growth in the city. (Personal communication, November 2016)

The city has the resources to measure air quality and traffic volumes at the street level. City officials are in a position to measure the results of their actions quickly. As an example, the toll charge at key points of entry and exit from the city centre had an immediate and measurable effect of reducing traffic by approximately 22%, which has been consistent over time (Eliasson 2014). The traffic in the inner city has been reduced by up to 15%, and this has contributed to a decrease in CO_2 emissions across the Stockholm County by 2–3% (ibid.). The reduction is considerable when considered in the light of continued population growth in the city. This is a source of considerable pride within the city government. One senior planner claimed:

> Council figures show that despite the population increase over the previous decade car trips in the city continue to decrease as commuters shift to public transport (80% at peak hour). We introduced the first increase in tolls in a decade in January 2016 and results show that further reductions in traffic volumes will be possible. (Personal communication, November 2016)

Mainstreaming and Incorporation

The city has promoted its ILS as the tool applied to coordinate policy implementation across functional areas. Evidence suggests that there have been political and administrative challenges in achieving comprehensive coordination of the climate *Action Plan*. One example is the coordination between the Environment and Health Committee and the Traffic and Waste Management Committee for the 'Travel by foot and by bicycle' objective, where the *Programme* specifies:

> Travel by bicycle should be doubled at the city's checkpoints as a result of the City's efforts. The proportion of pedestrian and bicycle travels should be increased to at least 20% in the suburbs and at least 50% in the inner city. (City of Stockholm 2012, 9)

Koglin (2015) describes that the power relations in Stockholm are not necessarily in favour of bicycle traffic (p. 220). The Traffic Committee has prioritised car-focused infrastructure to reduce congestion rather than the promotion of other forms of transport. One senior manager observed:

> The ILS has not been as effective as we hoped in overcoming political agendas in different committees across the council. For example the priority of our traffic committee has been focused on resolving traffic issues. This means roads come first in its agenda. Cycling infrastructure comes a distant second or third for traffic engineers. Cycle lanes tend to be retrofitted, somewhat reluctantly, long after the roads have been established or improved. This approach has obviously been unsatisfactory for the cycling objectives in the Action Plan. (Personal communication, November 2016)

Discussions with senior managers reveal that the city recognises these challenges and is implementing a new performance reporting regime to more clearly reveal the areas where attention needs to be directed. These managers remain optimistic that new processes and procedures for reporting will improve the coordination and achievement of the climate goals. In addition, the city has adopted project-based focus groups consisting of senior managers from relevant departments to work on individual developments such as the Royal Seaport. Staff argue that this approach, 'helps to identify potential problems early and to work on solutions together' (personal communication, November 2016).

Use

The city has been effective in using performance information to generate political support for climate initiatives. The city uses its performance information in a variety of ways to communicate with external stakeholders and to promote policy and management change within the administration. As an example, the city has established an Energy Centre to use the performance information it gathers through its energy efficiency programmes. Improvements in the heating and cooling performance of the city government's own buildings helps shape the information and advice provided to private building owners. Information can be used to demonstrate the effectiveness of its policies. The Council also uses its performance data to support participation in joint ventures with the private sector in demanding higher levels of energy-efficient performance in new constructions. Construction is an important policy area for the city government, as it is the largest provider of rental accommodation in the city. In 2016, there was a severe shortage of accommodation, with applicant waiting lists of up to 20 years (Savage 2016). The city plans to build 140,000 new apartments by 2030 and energy efficiency is a major objective. Being the lead partner in these arrangements helps ensure a continuous improvement in new constructions and acceptance within the private sector of higher levels of efficiency. The Sustainable Jarva project is a case of such a joint venture approach where the city maintains ownership of the land to ensure it retains influence over the development. The Council staff claim that policy learning has been taking place in the city, as senior officers argue that this approach has been the result of poor results in the past where private developers were able to resist requirements to improve the energy performance of buildings.

Management and policy change are supported through reporting of results of climate initiatives in different units across the Council administration—in areas where success can be reported and in areas where units fail to achieve their targets. One senior manager argued:

> The fact that our council is made up of minor parties and coalitions, all with their different agendas means we need to be ready to respond quickly with performance information relating to the actions we are taking. (Personal communication, November 2016)

Procedures established in 2016 will see changes to reporting so that the environment function is no longer amalgamated into the annual report to the Council. Senior managers in the Environment Department have successfully advocated for distinct voting, which will raise the awareness of at the political level and achieve greater support for the change in the implementation of climate policy:

> We need to improve the existing reporting of performance because the aggregated format through the annual report has not worked in terms of raising issues of poor performance. We need to apply improved reporting formats that are clearer and help hold individual sections to account more effectively. (Personal communication, November 2016)

On the broader community scale, communicating the performance message has been successful, as there is recognition for the work the city has been doing to respond the climate change. The regular community opinion surveys reveal a high level of awareness and support for the climate objectives (City of Stockholm 2013: personal communication, November 2016).

Conclusion

It is important to note that responding to climate change in Stockholm has not required dramatic departure from the policy and management practices of the past. Investment in renewable energy, public transport networks, and energy efficiency in building and construction systems represent an incremental change in the trajectory of policy decisions by the City Council over the twentieth century. Responding to climate change in Stockholm is starting from a strong base. The majority of climate initiatives are also built on a well-established working relationship and various forms of partnership between the city, other public sector agencies, the academia, and the business community. The city has a track record that is essential to develop these forms of partnership—particularly between the local and county authorities and national agencies in the Stockholm region. Shared and active planning in the past in fields of residential construction, transport infrastructure, and public transport systems provides a legacy that current decision-makers are continuing. An important lesson for other city governments is that the establishment of the ILS, with its corresponding performance framework, has been an attempt by the Council to ensure consistent approaches are taken when responding to climate change, particularly where partnerships are required.

The experience of Stockholm shows that broad political commitment from the Council is not necessarily producing the projected climate outcomes. The nature of politics in Sweden, with its coalitions of minor issue-based parties, creates opportunities for individual politicians to influence decision-making through the committee system. In Stockholm, this has been apparent in the Traffic and Waste Management Committee, where political representatives have been a block to advancing cycling infrastructure in the city. Managers in the EHA concede that the process established by the ILS often fails to produce the necessary coordination and achievement of the goals. This is an important lesson for other cities, as some of the poor-performing units are able to avoid exposure and accountability in contributing to the targets. The solutions being undertaken are an indication of the challenges the city government is attempting to overcome. Senior managers claim that the new reporting formats will require more itemised reporting against the climate and environmental goals. In future, individual units will report whether they achieved their agreed performance targets. Part of the reporting will require details on why they did or did not achieve their target, what resources they need to achieve the goal in the next reporting period, and any suggested improvements they could make, in addition to the original goal, that would exceed the target (personal communication, November 2016). The review of the *Action Plan* in 2016 has re-emphasised internal and external stakeholder support, with commitment to reduce the time frame for achieving the fossil fuel–free goal by 10 years, from 2050 to 2040. Whether the strengthened political commitment to achieve climate goals, combined with the improved management and reporting, will be successful remains to be seen.

References

Åström, D. O., Forsberg, B., Ebi, K. L., & Rocklöv, J. (2013). Attributing Mortality from Extreme Temperatures to Climate Change in Stockholm, Sweden. *Nature Climate Change, 3*(12), 1050–1054.

Birch, H. (2015). Where Is the World's Greenest City? *The Guardian*, Thursday, April 2.

C40. (2014). *Global Protocol for Community Scale Greenhouse Gas Emission Inventories.* http://www.ghgprotocol.org/city-accounting

C40. (2015). Cities 100: 100 Solutions for Climate Action in Cities. http://www.sustainia.me/cities/

City of Stockholm. (2007). *Adapting to Climate Change in Stockholm: Stockholm's Action Programme on Climate Change.* Environment and Health Administration.

City of Stockholm. (2011). *Stockholm Action Plan for Climate and Energy 2012–2015: With an Outlook to 2030.* Environment and Health Administration.

City of Stockholm. (2012). *The Stockholm Environment Programme 2012–2015.* http://international.stockholm.se/globalassets/ovriga-bilder-och-filer/the-stockholm-environment-programme-2012-2015.pdf

City of Stockholm. (2012a). *City of Stockholm Annual Report 2012.* www.stockholm.se/PageFiles/452199/city_of_stockholm_annual_report_2012.pdf

City of Stockholm. (2013). Waste Management Plan for Stockholm 2013–2016: On the Way to a World-Class Stockholm. http://www.stockholmvattenochavfall.se/globalassets/pdf1/riktlinjer/avfall/avfallsplan/avfallsplanen_eng_webb.pdf

City of Stockholm. (2014). *City of Stockholm Annual Report 2014*. http://international.stockholm.se/globalassets/rapporter/annual-report-2014.pdf

City of Stockholm. (2015a). *The City of Stockholm's Annual Report 2015*.

City of Stockholm. (2015b). Vision 2040: A Stockholm for Everyone. http://international.stockholm.se/governance/vision-2040/

City of Stockholm. (2016). Strategy for Fossil Fuel free Stockholm by 2040. http://international.stockholm.se/globalassets/rapporter/strategy-for-a-fossil-fuel-free-stockholm-by-2040.pdf

City of Stockholm. (2017). City of Stockholm: Sustainable Efforts. http://international.stockholm.se/city-development/sustainable-efforts/

Colding, J. (2013). Local Assessment of Stockholm: Revisiting the Stockholm Urban Assessment. In *Urbanization, Biodiversity and Ecosystem Services: Challenges and Opportunities* (pp. 313–335). Dordrecht: Springer.

Dastur, A. (2005). *How Should Urban Planning Engage the Issue of Sustainable Development? The Case of Hammarby Sjostad, Stockholm*. Doctoral dissertation, Columbia University.

EC (European Commission). (2010). *European Green Capital Award*. http://ec.europa.eu/environment/europeangreencapital/winning-cities/2010-stockholm/

Eliasson, J. (2014). *The Stockholm Congestion Charge: An Overview*. CTS Working Paper 2014:7. Centre for Transport Studies, Stockholm.

Gunnarsson-Östling, U., Edvardsson Björnberg, K., & Finnveden, G. (2013). Using the Concept of Environmental Sustainability: Interpretations in Academia, Policy and Planning. In J. Metzger & O. Rader (Eds.), *Sustainable Stockholm: Exploring Urban Sustainability through The Lens of Europe's Greenest City* (pp. 51–70). New York, NY: Routledge.

Iverot, S. P., & Brandt, N. (2011). The Development of a Sustainable Urban District in Hammarby Sjöstad, Stockholm, Sweden? *Environment, Development and Sustainability, 13*(6), 1043–1064.

Koglin, T. (2015). Organisation Does Matter—Planning for Cycling in Stockholm and Copenhagen. *Transport Policy, 39*, 55–62.

LSE (London School of Economics). (2013). Stockholm: Green Leader Report: A Report by the Economics of Green Cities Programme at the London School of Economics and Political Science. http://www.globalurban.org/Stockholm_Green_Economy_Leader_Report.pdf

Mahzouni, A. (2015). The 'Policy Mix'for Sustainable Urban Transition: The City District of Hammarby Sjöstad in Stockholm. *Environmental Policy and Governance, 25*(4), 288–302.

Metzger, J., & Olsson, A. R. (2013). *Sustainable Stockholm: Exploring Urban Sustainability in Europe's Greenest City*. New York: Routledge.

Mörtberg, U., Haas, J., Zetterberg, A., Franklin, J. P., Jonsson, D., & Deal, B. (2013). Urban Ecosystems and Sustainable Urban Development—Analysing and Assessing Interacting Systems in the Stockholm Region. *Urban Ecosystems, 16*(4), 763–782.

Norton, A. (1994). *International Handbook of Local and Regional Government*. Cheltenham: Edward Elgar.

OECD. (2013a). *Green Growth in Stockholm, Sweden*. OECD Green Growth Studies, OECD Publishing. DOI 10.1787/9789264195158-en

OECD. (2013b). *Water and Climate Change Adaptation: Policies to Navigate Uncharted Waters,* OECD Studies on Water, OECD Publishing. DOI 10.1787/9789264200449-en

Rutherford, J. (2014). The Vicissitudes of Energy and Climate Policy in Stockholm: Politics, Materiality and Transition. *Urban Studies, 51*(7), 1449–1470.

Savage, M. (2016). The City with a 20yr Waiting List for Rental Homes. *BBC News*. http://www.bbc.com/capital/story/20160517-this-is-one-city-where-youll-never-find-a-home

Stockholm County Council. (2013). Regional Planning in the Stockholm Region. https://www.uudenmaanliitto.fi/files/17164/Regional_planning_in_the_Stockholm_Region.pdf

Stockholm Stadus AB. (2015). Annual Report. http://stadshusab.stockholm.se/globalassets/stadshus-ab-annual-report-2015.pdf

Svane, Ö., Wangel, J., Engberg, L. A., & Palm, J. (2011). Compromise and Learning when Negotiating Sustainabilities: The Brownfield Development of Hammarby Sjöstad, Stockholm. *International Journal of Urban Sustainable Development, 3*(2), 141–155.

Sweden (Government of Sweden). (2007). Sweden Facing Climate Change—Threats and Opportunities. Final Report from the Swedish Commission on Climate Vulnerability. Stockholm. Swedish Government Official Reports: SOU 2007:60.

Sweden (Government of Sweden). (2014). *Sweden's Sixth National Communication on Climate Change*. Ministry of Environment, Sweden. https://unfccc.int/files/national_reports/annex_i_natcom/submitted_natcom/application/pdf/swe_nc6_resubmission.pdf

The Economist. (2012). *The Green City Index: A Summary of the Green Index Research Series*. Economic Intelligence Unit. https://www.siemens.com/entry/cc/features/greencityindex_international/all/en/pdf/gci_report_summary.pdf

Wollmann, H. (2004). Local Government Reforms in Great Britain, Sweden, Germany and France: Between Multi-function and Single-Purpose Organisations. *Local Government Studies, 30*(4), 639–665.

6

Copenhagen City Government

Introduction

The Lord Mayor of Copenhagen wanted his city to be the world's first carbon-neutral capital by 2025 (City of Copenhagen 2012a). The Government of Denmark also wanted Denmark to be the showcase to the rest of the world that a 'green transition can be reconciled with economic growth' (DMCEB 2013, 10). There has been strong focus on improving economic performance. Government figures reveal that Denmark has been slower than other Scandinavian countries to improve its gross domestic product (GDP) growth since the 2008 financial crisis (Levering 2014; The Danish Government 2016). Climate change has been framed as an opportunity to capitalise on both Denmark's and Copenhagen's international competitive advantages. The shared political view is that Copenhagen is the economic driving engine of Denmark (Capital Region of Denmark 2015). Dealing with climate change in the city has been placed in this economic context as a means of providing the co-benefits of new opportunities and making Copenhagen a more attractive place to live and invest. It is the combination of environmental and economic initiatives that the city argues will benefit everyone. The Mayor claims that Copenhageners have much to gain from the implementation of a set of policies outlined in a city

S. Jones, *Cities Responding to Climate Change*,
DOI 10.1007/978-3-319-64810-1_6

government climate plan, *CPH 2025* (City of Copenhagen 2012b). His argument is framed in economic terms when he argues that 'with the Climate Plan, we invest in growth and quality of life: clean air, less noise and a green city will improve everyday life for Copenhageners. The investments will secure jobs here and now—and the new solutions will create the foundation for a strong, green sector' (City of Copenhagen 2012a, 4).

Copenhagen has been recognised through international awards for its green credentials and attractiveness for investment. The OECD argues that 'Copenhagen has many lessons to offer other cities and local areas seeking to begin or expand their low carbon transition' (2012a, 101). In 2014, Copenhagen was the European Commission's Green Capital because it was a 'highly successful role model for the green economy, with an efficient communication strategy and the commitment required to develop its role as a model for Europe and beyond' (EC 2014). In 2015, the World Bank nominated Denmark as the best place in Europe to do business for the fifth consecutive year. Factors giving Denmark the advantage included a highly skilled workforce, innovative companies and industry clusters, and very competitive taxes and business costs (World Bank 2016). In 2013 and 2014, the international *Monocle* magazine, which popularises the notion of 'creative cities', voted Copenhagen the 'most liveable city in the world' (Monocle 2014). Success is attributed to the city government's ability to 'combine liveability, sustainable development and creativity' (City of Copenhagen 2015b). It is this approach that is at the heart of Copenhagen's *CPH 2025*. The Mayor believed that mega-cities on all continents were watching Copenhagen when it came to climate protection, CO_2 reduction, sustainable urban development, and green growth. One senior manager from the city government explained:

> Over 400 delegations from city governments across the world have visited the city to discuss Copenhagen's climate initiatives—we have staff whose sole function is to look after these delegations. (Personal communication, October 2016)

In 2014, the Lord Mayor was appointed to the steering committee of the C40 in recognition of the considerable efforts made by the city and the lessons it provided for other cities. Of particular interest to other city governments has been the transition to a green economy and efforts to address climate change in collaboration with local stakeholders (City of Copenhagen 2014a).

CPH 2025 reflects a performance-based approach to policy implementation. The climate plan outlines specific targets against policy objectives of the city government to reduce GHG emissions. The plan also provides details of how targets will be achieved, including the specific actions to be taken. These actions are subject to public performance reports that are updated on a regular basis. Recent reports are suggesting that achieving the objectives outlined in *CPH 2025* is proving to be increasingly difficult for the city government. In 2015, the city reported that it was falling behind in the goal for carbon neutrality by 2025, improving energy efficiency in existing buildings was harder than expected, and there were lower-than-predicted emission reductions from biomass conversions in the community heating system (City of Copenhagen 2015a, 6). In addition, the national government failed to introduce a congestion charge for Copenhagen in 2012 and dropped tax exemptions on electric vehicles in 2015; there was also reduced investment in climate initiatives, with budget cuts in 2015 to development and demonstration programmes within the areas of technology, environment, and green business development (Levering 2015a, b).

In Denmark, metropolitan governments must find alliances with relevant stakeholders if they are to successfully achieve their policy objectives. Successive reforms to the Danish local government since the 1990s have contributed to a new agenda focused on growth. Reforms in 2007 meant that Copenhagen lost several of its previous responsibilities and became a municipality without any special rights and responsibilities above other municipalities, despite being the capital and so important to the economy of Denmark (OECD 2009). In response to the reforms, the city government has adopted increasingly entrepreneurial forms of organisation and management (Hansen 2009). It has adopted market-driven approaches to form the foundation for Copenhagen's climate change strategies. The approach has also been the source of critiques that question the real objectives behind the policy action being taken by the city (Bayliss 2007; Lund Hansen et al. 2001).

Climate Change in Copenhagen

The national Ministry of Climate, Energy and Building (the Ministry) provides an outline of the long-term impacts of global warming on Denmark that also has relevance to Copenhagen. The city government has undertaken assessments, largely based on IPCC data, of how climate change will have direct impacts on the city. The reports provide relevant information and a focus for policy action by both governments. Both reports make projections to 2100 in a range of areas but primarily focus on temperature increases, increased flooding, and extreme weather events. In terms of environmental impacts, more precipitation, higher temperatures with a greater number of and more intense heat waves, and rises in sea level will increase the pressure on Copenhagen's biodiversity. The increased quantities of rain increase the risk of the sewer system overflowing. This may mean that the city's rivers, lakes, and coastal waters are polluted, and that biological balance may be disturbed (City of Copenhagen 2011, 55; 2012a). Climate change will also have social and economic impacts in Copenhagen. More specific details on the impacts are covered in the following sections.

Rising Temperature and Heat Island Effect

Future climatic conditions pose new challenges for Copenhagen. Temperature rises of 2–3°C, summers with longer periods of drought, and a greater number of and more intense heat waves are all predicted to occur. These heat waves, combined with the 'urban heat island' phenomenon in urban areas, may lead to a reduced quality of life for the population of Copenhagen and may result in increased public expenditure on energy consumption and health (City of Copenhagen 2011, 41). Future global warming will lead to Denmark warming in line with the trend already observed. The Ministry predicts that by 2050, temperatures will increase by around 0.9°C in the summer and 1.5°C in the winter. Towards 2100, temperatures are expected to increase 1.5–2.6°C in the summer and 2.3–3.8°C in the winter, depending on the emission scenario

(DMCEB 2013, 245). Recent assessments confirm that the 2100 scenario may have already been reached, with temperatures in 2015 already 1.3°C above average (Lai 2017).

The city government has been concerned that rising temperatures will contribute to the formation of urban heat islands in the city; in recent years, there have been wide regional variations, with a clear trend towards higher surface temperatures in built-up areas, including inner Copenhagen. These studies show relatively wide fluctuations in surface temperatures and posit that the urban heat island effect will cause major problems (City of Copenhagen 2011, 42). The hottest days have seen surface temperatures of up to 47°C, with some consecutive days between 32°C and 44°C (City of Copenhagen 2011). Surface temperatures in Copenhagen differ widely, with dense and high-building development areas, large covered areas, and, in periods of drought, large areas of grass also affected by heat.

Precipitation and Sea Level Rise

Ministry assessments suggest that in Denmark by 2100, there will be a tendency towards increased precipitation, with the largest increase of 17–27% in the winter season (Government on Denmark 2013). For spring and autumn, precipitation is also expected to increase, but the summer precipitation will be characterised by longer dry spells and more heavy precipitation events. For Copenhagen, the city government predicts that the intensity of rain is expected to increase 40% by 2100 (City of Copenhagen 2011). The rise in rainfall and the increased likelihood of storm surges combine as the dominant factors causing flooding in parts of the city. Flooding has been assessed in terms of its economic cost to the city. In the adaptation plan, the city government claims that damage from flooding cost the government DKK (Danish Krone) 350 million per year. Without climate adaptation measures, this figure rises to DKK 570 million in 2060 and DKK 1 billion in 2110 (ibid.). Flooding based on 100-year rain events already causes damage to 742 hectares across the city. The then city planner argued that 'for Copenhagen, the most serious effect of climate change will be increased precipitation, so we've developed a plan that addresses how to catch all the rainwater in the city' (Braw 2013). The Ministry report shows that 'towards the year 2100, the sea level around Denmark is expected to increase by approximately 1.5 m' (DMCEB 2013; 246). Significant changes need to be made to the storm-water system throughout the city, considering storm surges would also be combined with the possible sea level rise. For Copenhagen, a 400-year storm-surge event of 1.7 m today is expected to occur every 1–2 years towards the end of the century.

Extreme Events

Copenhagen has been experiencing extreme weather events that the city government attributes to climate change. Heavy rains in 2011 and 2014 caused considerable damage across the city. Computer modelling shows that an increased greenhouse effect results in higher frequency, increased intensity, and longer duration of extreme weather events. The national government expects Denmark to experience an increase in the frequency and duration of heat waves. Summers will be characterised by longer dry spells and an increase in heavy precipitation events. More extreme rain events are expected to result in increased flooding. Extreme events in the future are primarily expected to take place at the end of the summer.

Extreme precipitation determines the dimensions of wastewater systems. If the capacity of the drainage systems is exceeded, there is a risk of flooding. This may result in water damage to buildings. Torrential rain may also cause more basement floods as a result of more wastewater on the surface (City of Copenhagen 2011, 67). Even before the downpours in the summer of 2014, damage had been estimated at between DKK 6 and 9 billion. Conservative projections by the city government had estimated the total damage over 100 years at DKK 16 billion. The chief consultant for climate affairs at the City of Copenhagen argues that 'the citizens of Copenhagen are already feeling the effects of climate change. And there is an eagerness to get started on our journey towards adaptation' (City of Copenhagen 2014a, 34).

The Role, Responsibilities, and Trends in Local Government

The Constitution of Denmark of 1849, Paragraph 82 establishes 'municipalities' right to manage their own affairs independently, under state supervision, shall be laid down by statute'. Administrative–territorial reform of local government has been undertaken by the national government since the 1960s. An amalgamation agenda in 2007 resulted in the reduction of 14 counties to 5 regions, and 271 municipalities to 98. The municipalities and the regions governed by popularly elected councils are allowed to take on any task of their concern as long as it does not fall under the national government or another municipality/region. Furthermore, any administrative procedure must be in accordance with national laws.

The prevailing objectives of Denmark's reforms to the local government have been to make the local government system more efficient by providing a safeguard against the increasing future challenges—domestic and foreign. There are six key challenges Denmark aims to address:

1. maintain its ability to attract investments,
2. create good conditions for local economic development,
3. provide good education possibilities and an attractive environment for business managers and the local population to live in,
4. provide public services at a high quality for a low price,
5. represent Danish municipalities internationally, and
6. integrate international developments in the strategic development of municipalities so they remain competitive.

Reforms to the Danish local government over the last 40 years have resulted in the decentralisation of many national responsibilities to the local level. These changes have produced consequences for the management and administration of municipalities. Table 6.1 reveals the magnitude of the transition, with a reduction in the number of political representatives and a significant increase in the number of managers and administrators. In large measure, the municipalities implement national

Table 6.1 Municipal politicians and local administrators following local government reforms

	1966	1990	2010
Number of municipal politicians	10,005	4677	2468
Number of local administrators	46,020	451,916	527,775
Number of administrators per politician	4.6	96.6	213.8

Source: Blom-Hansen, (2012, 158)

programmes in a broad cross section of policy areas. The focus on the greening of Copenhagen's local economy rests on a long Danish tradition of advanced urban planning, as well as an early mover advantage within renewable and efficient energy technologies (OECD 2009).

Political representation is manifest through political parties and independents joining alliances. The largest coalition appoints the Mayor, who, apart from being head of the City Council, also functions as head of the municipal administration. Danish municipalities also appoint a city manager (CEO), on a permanent basis, who is responsible for all administrative matters, including working conditions, the structure of the administration, and the handling of cases.

Access to income tax, property tax, and corporate tax as sources of revenue is traditionally considered the foundation of the Danish local self-government. Table 6.1 shows the primary sources of revenue for municipalities in Denmark. The second most important income source is grants from the central government: matching grants to help finance local transfer payments and a general block grant. Municipalities can also levy fees in association with the supply of utility and social services. Borrowing is only permitted when exemptions are granted by the national government (Fig. 6.1).

Municipal functions amount to more than half of all public expenditure in Denmark. These functions include basic welfare services: childcare, primary education, and eldercare. In addition, municipalities have administered social transfers such as housing benefits, social security, and old-age pensions (Blom-Hansen 2012). Local authorities have responsibilities in the utility area, garbage collection, local roads, and in culture and recreation, libraries, theatres, and sports facilities. Observers argue that, since the municipal reforms of 2007, in terms of functions, 'the Danish municipalities today are stronger than ever' (ibid.). More than half the Copenhagen budget is spent on social services, such as employment benefits, eldercare, and childcare (OECD 2009, 210). The most important expenditure items on municipal budgets are education, social services, childcare, active labour market policy, culture, local roads, and utilities.

Urban planning has always played a large role for Denmark's municipalities and the reforms of 2007 strengthened this element of local responsibility. Prior to the reforms, spatial planning, environmental control, and water management were joint responsibilities of regions and municipalities (LGDK 2009). Following the 2007 reforms, these functions became the full responsibility of municipalities. Figure 6.2 illustrates

	Municipalities	
	Million DKK	Percentage
Taxes	233,106	52
Grants	145, 379	32
Fees	58,379	13
Net borrowing	2,316	1
Other	9,430	2
Total	448,651	100

Fig. 6.1 Revenue sources for local government in Denmark (2012)
Source: Blom-Hansen (2012, 51)

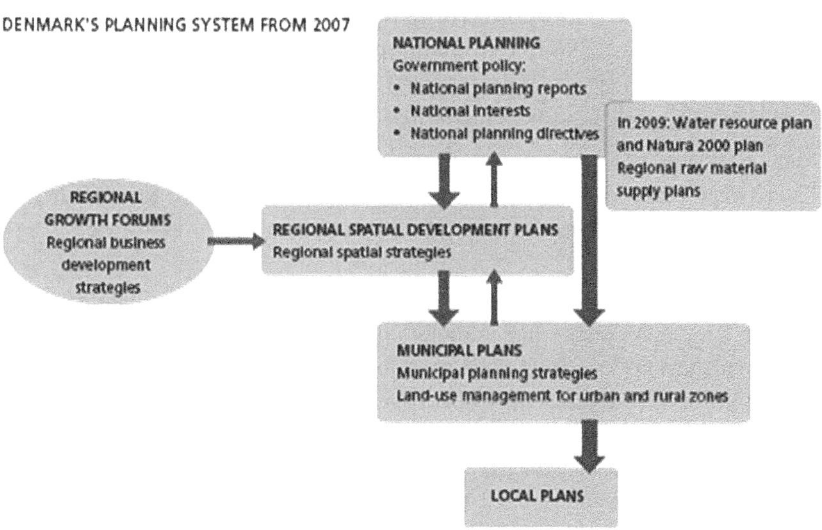

Fig. 6.2 Planning arrangements in Denmark between national and municipal levels

the relationships between national, regional, and municipal governments in terms of planning and the links between national planning and land use management plans at the city level.

Danish municipalities are required to convert the general planning guidelines of the national government into actual spatial planning. The municipal plan is prepared by the municipal council and provides a comprehensive and coordinated plan for land use in the towns and in the countryside. As a result of the structural reform, municipal planning will contain guidelines for many areas that were previously dealt with at regional level. This should be seen in line with the new responsibility for municipalities in relation to nature protection, environmental protection, and roads. It is important that the municipalities should reflect the national and international obligations regarding nature and ecological networks in their plans.

In 2007, the national government transferred employment services and the financial responsibility for unemployment benefits to municipalities. As a result, local councils face strong demands to promote employment opportunities within their jurisdictional boundaries. As the acknowledged driving engine of Denmark's economy, the challenge for Copenhagen to attract investment and create local job opportunities has intensified under these new arrangements. In 2016, Copenhagen's overall unemployment rate was 9–10%, regarded as high in the Nordic countries (Nordregio 2016). More concerning for the city's reputation as a creative hub has been the '25% unemployment rate among newly qualified professionals' (City of Copenhagen 2014b, 35). The city government is under pressure to establish initiatives that can produce results for the local economy, despite the fact that the national government retains the levers over macroeconomic policy. This is a serious issue for the city as the *Municipal Planning Strategy 2014* makes clear:

> More Copenhageners need to be in work. Current challenges include the 25% unemployment rate among newly qualified professionals in Copenhagen and the long-term need for more citizens to have vocational qualifications. In parts of Copenhagen, the proportion of citizens in employment is low, so we are working purposefully to bring the level of employment in disadvantaged areas of the city up to the Copenhagen average. (City of Copenhagen 2014b, 35)

The Copenhagen lifestyle and the green economic initiatives taken are seen by the city as key strengths that need to be exploited as part of the strategy to improve economic growth and increase employment opportunities. The *Municipal Plan 2015* makes this very clear and suggests that the attraction of 'creative class theory' espoused by Richard Florida (2008) forms the basis of the approach:

> Copenhagen's liveability and green profile remain key to our ability to attract more tourists, entrepreneurial environments, foreign businesses and highly educated people, which thereby maintains the high quality of the city. (City of Copenhagen 2014b, 35)

The local government in Denmark has been depicted as a duality and tension between historic participatory approaches empowering welfare obligations and a new form of urban authoritarianism that requires entrepreneurial market-driven approaches to promote growth (Andersen and Pløger 2007). These tensions form part of the challenges faced by the City of Copenhagen as it attempts to undertake its traditional functions and responsibilities as well as manage demands created by new responsibilities for employment and economic development. The development and implementation of climate strategies is at the crossroads of these competing demands, as they are an attempt to deal with the impacts of climate change on the environment and the community, while also promoting new opportunities as part of a 'green transition' to support economic objectives. The political work in the City of Copenhagen builds on traditions of openness, democracy, collaboration, and citizen participation. New arrangements across the Danish public sector, including municipalities, are increasingly based on the neoliberal assumptions that governments can achieve much better results when they cooperate with companies and organisations than they can alone. For the city government, this approach means that the municipality needs 'to step out of the authority role and become equal partners who solve the city's challenges together' (City of Copenhagen 2014a). For some observers however, when taken together, these tendencies reveal that economic growth has become the primary goal of urban policy, and that private sector interests are taking a larger role in their formation than in earlier years (Bayliss 2007).

Governance in the City of Copenhagen

Challenges from inside and outside the city impact on the success of climate change initiatives. Governance arrangements in the Copenhagen city government consist of 55 elected representatives operating through committee rule, with a City Council and seven function-based committees. In terms of climate change, both the Finance Committee and the Technical and Environment Committee (TEC) share responsibilities that relate to both climate mitigation and adaptation initiatives. Under the governance arrangements, the Finance Committee is responsible for

> [t]he *Urban Development Centre* which is responsible for the City's overall urban development, including housing policy and community planning, traffic planning, traffic procurement, efficient and cohesive crime prevention, growth, partnerships and international affairs, purchase and sale of properties and the relations with enterprises owned by the City of Copenhagen. (City of Copenhagen 2014c, 12)

The TEC is responsible for '*City Construction* that includes changes in the city's layout, and the *City development function which* includes the creation of strategic, cohesive and predictable frameworks for the city's development' (ibid., 32). More broadly, the TEC is responsible for environmental and climate activities, the development of the traffic area, the development of new urban areas, and the city's green areas (ibid.). Senior managers argue:

> The link with the Finance Committee and the TEC has been a critical factor in the implementation of the CPH2025, because we establish a budget commitment for our programmes and initiatives. (Personal communication, October 2016)

Political support from the then Lord Mayor was critical to establishing the Council's commitment to climate action. Under these governance arrangements, the committees make final decisions within their areas, which reduces the number of cases that must be submitted to the City Council (City of Copenhagen 2014). The benefits of these arrangements, in combination with both political and budget commitment, senior managers believe, is that they have a clear mandate to implement *CPH 2025* climate initiatives (personal communication, October 2016).

According to the OECD, Copenhagen's committee-based system is not well-adapted to integrated planning and policy execution. The national government mandates municipalities to establish a committee-based governance system. The number and the subject area of committees are organised by each city, and the City Council determines the composition of the committees. The result of this governance model is a bottleneck largely restricting the flow of performance information within the organisation to vertical lines of control and putting strains on cross-organisational communication, although many of the areas of local government responsibility are cross-sectoral in nature. Council managers recognise this issue and the constraints it places on implementing climate policy.

Local politics in Copenhagen has strong social democratic roots. Decision-making in the city government has traditionally focused on community strategies and has been participatory, 'founded on priorities of radical democracy, social justice, inclusion and citizens' empowerment' (Andersen and Pløger 2007, 1349). Without strong support from business groups and the national government, the city faces significant challenges in achieving its climate objectives. The city has placed a strong reliance on private sector investment in its attempts to promote Copenhagen as a 'green' capital. The Lord Mayor faces challenges within the City Council based on a governance model the OECD describes as 'problematic' in terms of policy implementation (2009, 230). These internal and external challenges threaten to undermine the capacity of the city government to capitalise on what it calls its competitive advantages and to achieve the green transition the city has been planning for two decades (OECD 2009; LSE 2014).

Copenhagen City Government Climate Change Strategy

Climate change action in Copenhagen is detailed in various planning documents relevant to both mitigation and adaptation. Mitigation initiatives are focused on four themes: energy consumption, energy production, mobility, and changes to various aspects of city administration. These themes make up the approach taken by the city's guiding climate change planning document: *CPH 2025*. Where relevant, there are connections to climate initiatives across each of the municipality's guiding policy documents. Table 6.2 provides details of the documents that influence how the city government will take action on climate change. The city acknowledges that it is limited in its policy options because of the central role of the national government in providing the legislative framework for what it calls the 'green transition' of the city.

Climate adaptation initiatives responding to the climate risks such as flooding, temperature rise, and extreme events are outlined in the 2011 *Copenhagen Climate Adaptation Plan*. The plan is a high-level strategic document rather than a plan with detailed specific actions to be taken. The likelihood of progress being made will be dependent on 'future actions delivered through existing municipal planning and management functions' (LSE 2014, 99). Planning and actions taken by Copenhagen to use public space for adaptation initiatives may be constrained by national priorities. Proposals that use parks for climate adaptation purposes will need to follow a hierarchy of authority that extends from national statutory legislation down to planning and implementation at the regional and the local level (LSE 2014, 80).

In many respects, the Danish government and the city government have been complying with the obligations of membership of the EU. This membership impacts on emissions reduction targets and incentives to support the transition away from fossil fuels to renewable sources of energy. Local Government Denmark (LGDK) analysis shows that the EU affects approximately 50% of agenda items in Danish municipal councils (KL 2015, 1). Legal requirements established by the EU influence municipal decision-making particularly in terms of compliance with environmental standards, procedures for environmental assessment, and water quality. Figure 6.3 shows that the political influence of the EU on climate change policy at the municipal level impacts on approximately 30% of the initiatives taken.

Despite these constraints, Copenhagen has some confidence that, due to its size and position as a capital, it can try to influence the framework at the national level in areas of interest such as 'CO_2 reduction, energy

Table 6.2 Guiding policies of the Copenhagen city government for climate change

Danish Climate Policy Plan: Towards a Low-Carbon Society 2013	The Danish government's target is to reduce total Danish greenhouse gas emissions by 40% by 2020 compared with the 1990 level
How to Manage Cloudburst and Rainwater: Action Plan for a Climate-Proof Denmark 2012	Five areas of initiative: 1. Improved framework for climate change adaptation 2. More consultancy and new knowledge 3. Strengthening of collaboration 4. Green transition 5. International focus
Our Future Energy	The Danish government targets to secure 100% renewable energy by 2050.
Climate Strategy for the Capital Region of Denmark 2012	By 2025, the Capital Region will be the most climate-ready and energy-efficient region in Denmark based on strong regional and cross-municipality collaboration, where innovative public–private partnerships contribute to green growth of first-rate international calibre.
City of Copenhagen Municipal Plan 2015: A Coherent City	An environmentally sustainable focus on urban development supports the use of public transport and cycling, environmentally friendly energy supply, and common solutions for waste management, while preventing urban sprawl into surrounding open areas and green areas in the city.
Copenhagen Climate Adaptation Plan 2011	Focus on the following initiatives: 1. Development of methods to discharge rainwater during heavy downpours 2. Establishment of green solutions to reduce the risk of flooding 3. Increased use of passive cooling of buildings 4. Protection against flooding from the sea
CPH 2025 Climate Plan	The City of Copenhagen's plan for achieving carbon neutrality by 2025, with a focus on four areas: • Energy production • Energy consumption • Mobility • City administration

(*continued*)

Table 6.2 (continued)

Good, Better, Best: The City of Copenhagen's Bicycle Strategy 2011–2025	By 2015, at least 50% of people to cycle to their workplace or educational institution in Copenhagen. The number of seriously injured cyclists in Copenhagen to be halved compared with 2005 (when there were 118 serious injuries). At least 80% of cyclists in Copenhagen to feel safe and secure in traffic.
ITS Action Plan 2015–2016: Better Mobility in Copenhagen	Cyclists' average speed increased from 15–17 kph (kilometre per hour) and number of stops reduced by 10%. Pedestrians' improved transitions at intersections. Buses' average journey times reduced by 5–20% during peak hours, with a 10% increase in reliability of travel time. Cars' average travel time reduced by 5%, a 10% improvement in reliability of travel time at peak hour, and the number of stops reduced by 10%.

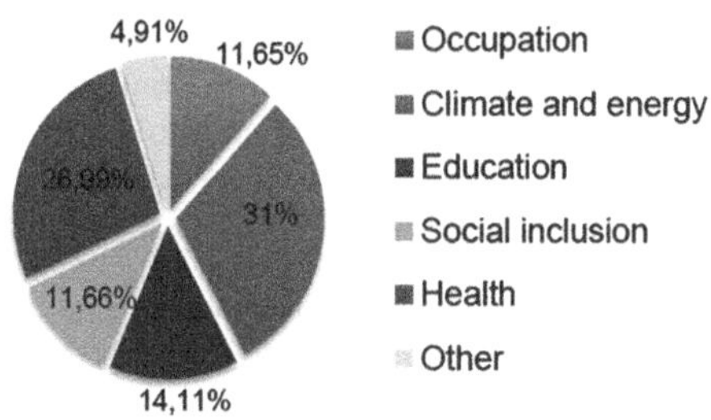

Fig. 6.3 Areas under EU's political impact
Source: Adapted from KL (2015)

production, energy consumption, retrofitting and new build, green transport, urban development, green growth and job creation' (City of Copenhagen 2012b, 12). The city government argues that the competitive international context means that Copenhagen needs to be more ambitious in its climate objectives in order to attract foreign investment to support the green economy transition necessary to improve economic conditions in the city (personal communication, October 2016). Ambition however has not been sufficient to overcome the constraints imposed by national decisions. The city experienced the impact of the national policies in 2012 when the national parliament refused to endorse a congestion charge in the city. This single action would be detrimental to the mitigation objectives of *CPH 2025*, as the congestion charge was predicted to reduce CO_2 emissions in the city by approximately 70,000 tonnes. In order to meet its own goal of carbon neutrality by 2025, the city has been forced to look for other areas where emissions could be reduced by, for example, expanding investment in wind energy (City of Copenhagen 2012b, 9). According to one senior manager:

> [T]he evaluation of CPH 2025 being undertaken in late 2016 shows the city remains committed to the overall carbon neutral objective but has realigned specific initiatives to make up for the deficit caused by the failure of the national government to support the application of a congestion charge for the city. (Personal communication, October 2016)

Closer examination of Copenhagen's approach to mitigation serves to highlight the linkages across the other planning documents, suggesting some form of mainstreaming of climate and sustainability considerations across council functions and a level of collaboration agreed in the original consultations that takes place through the policy implementation process. According to the city government, *CPH 2025* is 'constantly undergoing assessment and evaluation in order to stay on track' (City of Copenhagen 2012b, 21). This provides us with the first indication that some form of performance management framework is present and performance information is being used in the implementation of climate initiatives. Performance reporting is available primarily through a series of *Green Accounts* reports, produced in 2012 and 2014. Additional reports, both general and specific, have appeared, for example, *Copenhagen Climate Projects: Annual Report for 2014 & 2015* and *Copenhagen European Green Capital 2014: A Review*.

Many of the goals and initiatives adopted by the city reflect the national climate objectives and EU agreements supported by Denmark. Table 6.2 outlines the range of climate plans at the national level that serve as guides for local action. At the national level, the *Danish Climate Policy Plan: Towards a Low-Carbon Society 2013* contains details on the regulatory approach taken for mitigating emissions for Denmark. The EU and its Emissions Trading Scheme, which provides a set of requirements for member states, impacts on Danish internal energy consumption through CO_2 taxes and a price on CO_2 emissions. The Danish government claims that these taxes have contributed to 'reducing energy consumption and thus reducing carbon emissions in Denmark' (Government of Denmark 2013, 36). The national level also sets standards for maximum energy consumption in buildings and provides subsidy schemes to promote renewable energy. The goal of the Danish plan is to eliminate CO_2 emissions from the energy sector by ensuring the use of 'coal will have been phased out, while electricity and heat production is to be 100% covered by renewables by 2035. Fulfilling these goals will also eliminate CO_2 emissions from this part of the energy sector' (ibid., 37). Other EU regulations supported by Denmark that are of direct relevance to emissions in Copenhagen relate to performance standards for motor vehicles. These regulations support emissions reductions from cars through a shift to biofuels and the promotion of electric vehicles.

At the regional level, the Capital Region has established a *Climate Strategy for the Capital Region of Denmark 2012* that encourages cooperation and coordination by municipalities covering the City of Copenhagen and the surrounding 24 local governments. The main focus of the Capital Region is the establishment of adaptation plans. While the region has no legislative power in this area, it serves as a coordinator and facilitator to ensure there are linkages between the municipalities, residents, businesses, and knowledge institutions in areas where there are common interests and shared problems. According to the Capital Region, the climate strategy is part of the broader Regional Development Plan in relation to sustainability and climate. There is little capacity for follow-up because municipalities have no accountability or reporting requirements at the regional level (Capital Region 2012).

The *City of Copenhagen Municipal Plan 2015: A Coherent City* focuses on 'strengthening the quality of life for the Copenhageners and creating growth' (City of Copenhagen 2014b, 1). This links to the overarching goals for the Copenhagen region to maximise economic growth and innovation (Capital Region 2015). The *Municipal Plan* objective is to maintain and improve the high quality of life of Copenhagen residents

(City of Copenhagen 2014b), and to further improve the city's environmental performance by making Copenhagen the world's first carbon-neutral capital by 2025 (City of Copenhagen 2012b). These high-level objectives form part of spatial land use and transport coordination in the *Municipal Plan*. In specific connection to *CPH 2025*, the *Municipal Plan* reiterates the city's commitment to future directions for mobility in the city. Initiatives include distribution of traffic in Copenhagen, with at least 30% by bicycle, at least 30% by public transport, and no more than 30% by car. At least 75% of the growth in traffic must be green and at least 50% of traffic to work or school in the City of Copenhagen must be by bicycle.

An examination of the initiatives under the *Mobility* theme will be used to highlight the specific application of performance management within the Copenhagen city government. *Mobility* provides examples of direct local responsibility where the city government has power and resources to make considerable impact on GHG emissions. It is also an area of significant complexity, as it requires considerable behaviour change by both internal and external stakeholders. The examples illustrate the capacity of the city government to implement climate policy and help to illustrate how performance management has been applied throughout the process.

Mobility

According to data provided by the city government, in 2010, 544,000 tonnes of CO_2 was emitted by the transport sector in Copenhagen, corresponding to 22% of the city's total CO_2 emissions. Road traffic alone was responsible for 70% of transport emissions. Mobility initiatives provide substantial prospects for the city to make a difference, with a direct impact on emissions. The local government in Denmark has direct responsibility for spatial planning and infrastructure provisions, giving the city greater control over transport corridors, including roads, bus transport, and bicycle lanes. The initiatives adopted by Copenhagen for transforming transport to greener options are largely investment driven (LSE 2014, 46). Cycling has become an iconic symbol of Copenhagen:

'[B]icycles define Copenhagen's cityscape' (Pearson 2007). There seems to be little doubt that the city is the world's leader in terms of the 50% modal share of bicycles in the transport mix (Cathcart-Keays 2016). A key question for other city governments relates to the approaches taken by Copenhagen and why it has succeeded in increasing the numbers of drivers who change from cars to bikes. The city argues that there are sound economic reasons for promoting cycling. The argument is that increasing cycling will save time and money. A report commissioned by the city in 2011 reveals important insights into why cycling has become the prime 'mobility' focus for the local council (Sustainia 2011). Figure 6.4 shows how cycling has been analysed in terms of its value to the city in comparison with travel by car.

The report claims that, when looking at the total cost of 'air pollution, accidents, congestion, noise and wear and tear on infrastructure, when travelling by bike or car—bikes came out on top' (Sustainia 2011, 17). For every kilometre travelled by bike instead of car, the city saves approximately DKK 0.45 (ibid.). This evidence is often used by the Bicycle Secretariat to justify expenditure on cycling infrastructure (personal communication, October 2016). Promoting transport by bike and public transport, even linking the two, is seen as giving Copenhagen an international 'competitive advantage' in terms of investment attraction when compared with other cities. It is not only about reducing emissions, it is also about improving the quality of life and creating jobs and business

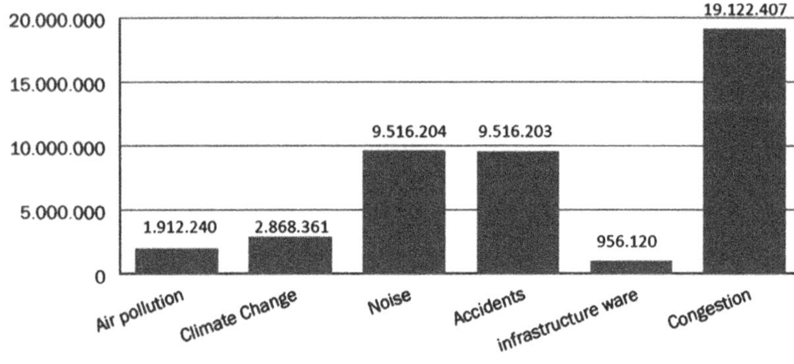

Fig. 6.4 Estimated avoided external costs, shift from cars to bicycles
Source: Sustainia (2011)

opportunities throughout the entire economy (ibid., 5). Of course these are vague and subjective notions that will defy accurate measurement, but they continue to be the ideal outcomes, as they are promoted by the city. The Bicycle Secretariat within the TEC develops arguments for investment in cycling in socio-economic terms, with cost–benefit analysis and 'return on investment' calculations forming the basis of submission to the Council for funding on an annual basis (personal communication, October 2016).

The 'City of Cyclists' mantra has been elaborated through the *Good, Better, Best: The City of Copenhagen's Bicycle Strategy 2011–2025*. This strategy aligns with the goals in *CPH 2025* by focusing on increasing the number of Copenhageners who choose to travel to work and school to 50%, improving and expanding cycling infrastructure, and improving safety for cyclists. These are aspects of the 'mobility' theme, over which the city has statutory and programmatic control. The other aspects, including public transport and alternative fuels for vehicles, will involve a 'mix of actors from the local to supra-national level for standard setting and emissions regulations' (LSE 2014, 97). The city has produced biannual cycling indicators since 1996 in the form of public reports—the *Bicycling Accounts*. These reports provide the city planners with rich data on cycling, both in terms of statistics and in terms of surveys of Copenhageners' perceptions of cycling conditions (Nielsen et al. 2013, 111). Figure 6.5 shows the actual results for the target objectives provided in the most recent report.

'04	'06	'08	'10	'12	'14	'15	'25	
36	36	37	35	36	45	50	50	**Proportion of people who cycle to work/education (%) ***
58	53	51	67	76	74	80	90	**Proportion of cycling Copenhageners who feel secure (%)***
125	97	121	92	102	91	56	34	**Cycling casualties (number per year)***
				17	19	40	80	**Proportion of PLUS network that has 3 lanes (%)**
				0	7	5	15	**Reduction in cycling travel time (%)**
50	48	54	50	61	63	70	80	**Satisfaction with state of cycle tracks (%)**
			67	73	70	70	80	**Satisfaction with cycling culture's impact on urban life (%)**

Fig. 6.5 Targets and results for cycling in Copenhagen 2004–2014
Source: City of Copenhagen (2014)

In the context of all this encouragement, one of the most interesting reporting measures in the *Bicycle Accounts* is the assessment of why Copenhageners cycle. It seems the residents have also adopted practical and economic reasons used by the city government. Surveys show that 50% of residents cite convenience as the prime reason for cycling; it is simply easier and faster to get around Copenhagen by bike. Research by the city government in 2011 attempted to identify reasons why 32% Copenhageners preferred to drive for short trips of approximately 500 m (see Box 6.1). The famous 'Finger Plan' spatial layout has promoted urban growth along rail corridors radiating from the city centre, combined with the limited road space in the centre is intended to make travelling by bicycle less challenging than by car. Other factors influencing the need for a car include the 'Station Proximity Principle', which generally requires new large offices of more than 1500 m^2 to be located within 600 m of a railway station, and city-level land use planning, which stimulates mixed-use, high-density development around stations and limits parking provisions (LSE 2014). The combination of these factors shape what some call the 'ordinariness' of cycling in Copenhagen (Freudendal-Pedersen 2015). Figure 6.6 reveals that unfortunately, in terms of environmental sustainability and climate change, only 7% of those who cycle have environmental objectives in mind (City of Copenhagen 2014a). Such figures provide a challenge for those city government advocates who use Copenhagen as an example of an environmentally focused city, where residents change their behaviour in order to reduce their impact on the environment. In this regard, the experience of Copenhagen is not readily transferable to other city contexts. Perhaps, surprisingly for advocates of cycling in other cities, the growing number of cyclists in Copenhagen is regarded as a negative feature by non-cyclists, and in particular pedestrians (Nielsen et al. 2013). The Danish media raises questions relating to the side effects of cycling promotion, for example, perceptions of safety and the social atmosphere of public spaces, which may be challenged by a policy that prioritises large numbers of cyclists.

In other aspects of the mobility theme, policy initiatives introduced by the national government have aligned with the plans of the city government. Policy initiatives have been introduced to encourage fuel switching. The increase of biofuels in the energy mix has been strengthened

Box 6.1 Short car trips in Copenhagen

Danish National Travel Survey data in 2010 revealed that 32% of car trips in Copenhagen city were under 5 km. Drivers within this cohort include parents driving their children to and from school, those driving to local shops, and inner-city dwellers driving to work. Short-distance drivers are of interest to city governments internationally, as they are a major contributor to traffic congestion in areas where there are alternative modes of transport easily available. Finding ways to convince these drivers to walk, ride, or take public transport will provide opportunities to improve traffic flow and reduce infrastructure costs in inner-city areas.

In 2011, the Bicycle Secretariat within the Copenhagen city government decided to investigate reasons why drivers preferred their car to a bike for these trips. The Secretariat took a four-pronged approach to investigate how they could more effectively communicate with these drivers and encourage the use of bikes as a preferred alternative. First, they initiated an extensive collection on data concerning drivers of short car trips. This included both qualitative studies such as social anthropologists who were employed to interview drivers, sit with them in their cars, discuss driving with their families, and discuss the possibility of riding a bike instead. Also focus groups with drivers would focus on practical and cultural barriers to switch the short car trips to cycling or walking. For quantitative inputs, a large segmentation study based on national travel survey data was carried out, as well as a broad survey of drivers to identify reasons for taking short trips.

Data showed that 30% of all drivers of short trips did not have access to a functioning bike. To help drivers overcome this practical barrier to take up more cycling, the city initiated bicycle repair service events in urban development areas where bicycle repair shops had not established yet and where it would take a 2 km walk with a broken bike to have it fixed. Citizens proved very happy with the scheme, and afterwards, several mobile bike repair start-ups now offer their services in the neighbourhoods.

Other data showed that 19% of drivers found that an online bicycle route planner might motivate them to cycle more, and that bicycle routes completely separated from car traffic were their major preference for infrastructure. Therefore, the city developed a free online route planner, *I bike CPH*, and integrated an easy way of finding green cycle routes and other green bicycle connections. This was combined with a general campaign to promote green cycle ways and an updated map showing where to find green cycle connections.

Juxtaposing these initiatives, the Secretariat worked with the Council's 'safe routes to school' programme, including educational programmes in after-school centres and the 'Mini Copenhageners' campaign, which generated

(*continued*)

Box 6.1 (continued)

interest and excitement among young riders through special identity plates for the bikes ridden to school.

By 2016, the work of the Secretariat had reduced the number of short car trips from 180,000 to 150,000, and the share of short car trips dropped from 32% to 25%. It is however difficult to isolate the effect of the soft measures alone, as investments in improving bicycle infrastructure were also intensified in this period. However, the results are an important lesson for other city governments, as they help to illustrate the power of well-researched data combined with an effective message. Such messages communicated to the right audience can produce behaviour change within the community without governments needing to take a harder top-down approach.

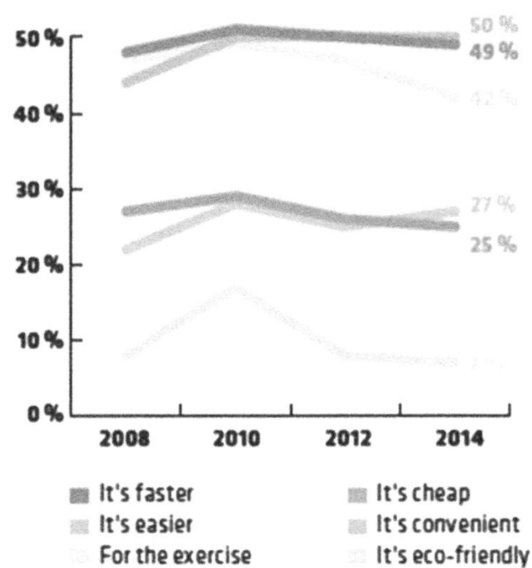

Fig. 6.6 Why Copenhageners cycle
Source: City of Copenhagen (2014)

through policies requiring all petrol and diesel sold in Denmark to contain 5.75% biofuels. An examination of the *Green Accounts* in Table 6.3 reveals that the city government has not pursued measurable activities to promote the transition to biofuels within the community. The city has instead focused its efforts on using these fuels in its own fleet of both cars and buses.

The city has also taken advantage of the subsidies available to purchase electric vehicles in its fleet. These national subsidies have reduced the cost of purchasing new vehicles. Denmark has encouraged the sale of electric vehicles by exempting buyers of fully electric powered vehicles from the current vehicle registration tax. The vehicle registration tax places a 180%

Table 6.3 Mobility initiatives in *CPH 2025*

Goal for 2025	Initiatives	Results
75% of all journeys in Copenhagen are on foot, by bicycle, or by public transport	Establish mobility programmes outlining green transport options to target groups	*Green Accounts* in 2012 and 2014 outline the following: • The second cycle superpath between Farum and Copenhagen was opened in April 2013. • In 2014, the Bicycle Snake opened to cyclists who make a shortcut by crossing the harbour between Vesterbro and Amager. • Targets have been set for the prioritisation of bicycles, pedestrians, buses, and cars on the various sections.
50% of all journeys to place of work or study are by bicycle	Developing bicycle connection in and to Copenhagen, partnerships for growth of cycling, improvements in conditions for cycling by promoting cycling workplaces	The first cycle superpath connecting municipalities surrounding Copenhagen with the city was opened in April 2012, and the second cycle superpath was opened in April 2013. So far, a grid of cycle superpaths covering a total of 300 km has been planned.
20% more passengers using public transport to 2009	Reduce bus travel times by 10%	Not reported
Public transport is carbon neutral	Increase the use of biofuels in buses, trial electric buses	In 2013, the city introduced two electric buses for a trial period to test whether this technology was compatible with public transport needs.

(*continued*)

Table 6.3 (continued)

Goal for 2025	Initiatives	Results
20–30% of all light vehicles use fuels such as electricity, hydrogen, biogas, or bioethanol	Demonstration projects to focus on biogas and bioethanol, establish secure supply and infrastructure for biofuels, establish parking and fuel stations for electric vehicles	To take the lead and show that there are good alternatives to fossil fuels, the municipality has invested in electric and hydrogen cars, and by the end of 2014, the city had 177 electric and hydrogen cars, which represent 53% of the total fleet.
30–40% of all heavy vehicles use new fuels	Establish secure supply and infrastructure for biofuels	Not reported In 2015, the city argued it still needed to push to replace trucks, buses, and cars running on fossil fuels with hydrogen, biogas, and electric vehicles. Contractors required to provide services to the Council with vehicles powered by biofuels

tax on the purchase of any vehicle after the first DKK 65,900 or $11,987 of the price of the vehicle (Berman 2013). Despite this deterrent, the take-up of electric vehicles remains comparatively low against other Scandinavian countries. Perhaps Copenhagen is a special case where the evidence on cycling shows the purchase of an electric vehicle would do little to improve travel convenience or reduce costs for travelling around the city. Research by McKinsey, illustrated in Fig. 6.7, reveals that countries such as Norway and the Netherlands have more effective incentives and have significantly increased the number of these vehicles when compared with Denmark. In contrast, the Danes have major concerns with electric vehicles, including range anxiety per charge, price compared with smaller economical vehicles, and lack of knowledge regarding benefits (Green et al. 2014). These factors however have not prevented Copenhagen from investing in electric and hydrogen cars. In 2015, 85% of all municipal light vehicles were running on alternative fuels (City of Copenhagen 2014d, 14).

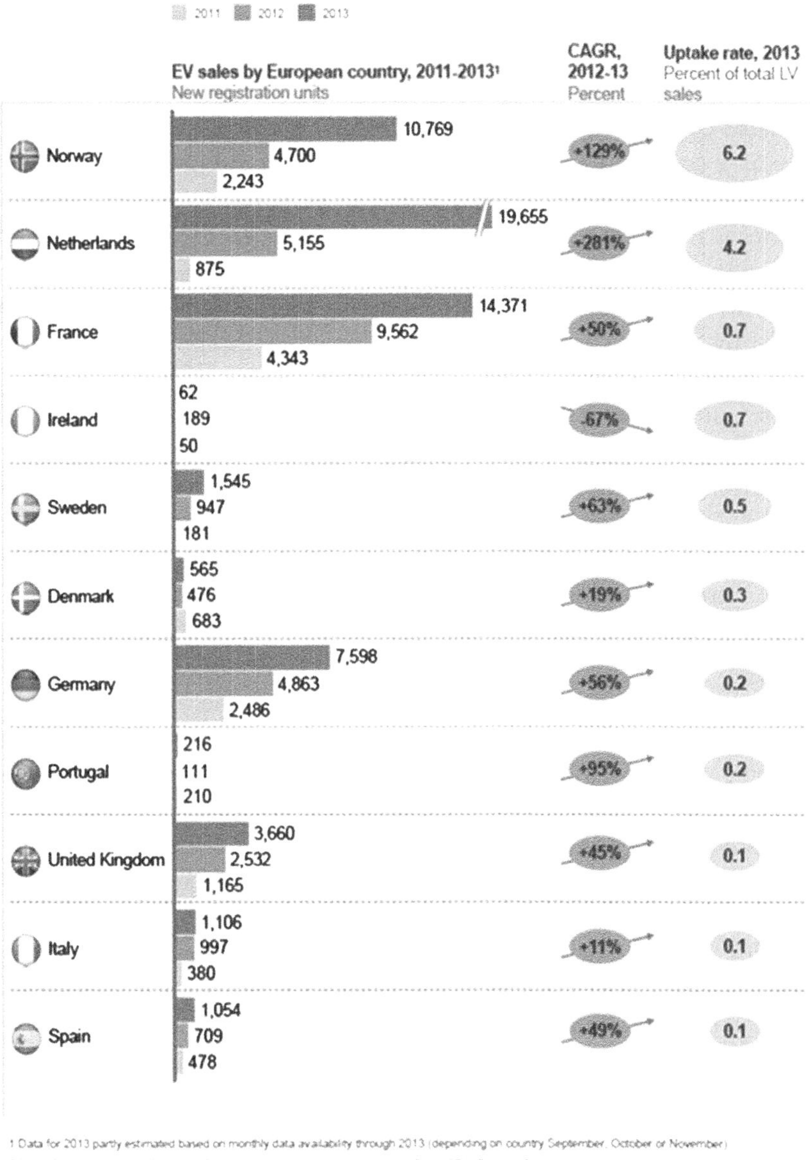

Fig. 6.7 Electric vehicle sales in Europe 2011–2013
Source: 'Electric Vehicle Sales and Subsidies' (McKinsey 2014)

The national election in 2015 saw a change of national policy that would remove tax incentives for electric vehicles. While Danes have not been convinced of the benefits of these vehicles, the city government remains committed to its goal of transferring the fleet to electric vehicles as a cost reduction measure. Perhaps the time for incentives will soon be over and will no longer be the preferred option for governments as they look for ways to reduce costs, lessen EU dependence on oil imports from Russia, and decrease emissions in major cities. The government of Norway is planning a ban on cars fuelled by petrol or diesel by 2025 (Staufenberg 2016). Several other countries in Europe, including Germany and France, are formulating similar programmes to phase out fossil fuel-powered transportation. In fact, Germany's (then) Deputy Economy Minister Rainer Bakke argued that 'there's been no reduction at all in CO_2 emissions by transport since 1990' (Govlink 2016). Once multimember states pass such a ban, there may continue to be attempts to enforce these rules across all member states. In 2017 there were ongoing efforts by national and city governments to demonstrate commitment to reducing emissions by promising to ban diesel and gasoline cars by 2040 (Ryan and Shankelman 2017).

The evidence provided by the *Mobility* theme shows that the Copenhagen city government has taken serious steps in a number of areas to apply a performance management framework when implementing its *CPH 2025*. Political commitment to the carbon-neutral goal has remained consistent despite the need to alter areas of focus resulting from constraints imposed by the national government. Senior managers and planners across the organisation in areas relevant to the implementation of the climate objectives are clear on what they need to achieve in approaching the outcome of carbon neutrality in the city. The political consistency and the budget commitments maintained over the first implementation period have been transferred to the second period following the first evaluation of *CPH 2025* initiatives after the first four years of help to maintain motivation and commitment of staff. The policy literature stresses the importance of this consistency as a major contributor to the implementation of policy initiatives. New approaches have been designed to compensate for the lack of support for the congestion charge and the reduced incentives for electric vehicles by the national government.

Leadership by the Lord Mayor and the TEC has been consistent since the development of *CPH 2025*. Such leadership has been shown in the literature to be critical to the implementation of climate policy at the municipal level. Without this support, climate change struggles to raise attention above more pressing issues across the political agenda. Part of the success of the Copenhagen approach has been the 'coupling' of the climate change agenda to the economic and employment agenda. The green economy is seen as the answer to the issues of the transformation of the local economy through investment attraction and job creation in the city.

Performance Management and the Copenhagen City Government *CPH 2025*

The city government argues that it takes a collaborative approach by consulting businesses, research units, Copenhageners, and the various city administrations. The Council states that 'without collaborating with relevant players about coordinating actions and implementing them, we will not achieve our goal' (City of Copenhagen 2012b). In establishing CPH2025, the Council's TEC followed an extensive 12-month consultation programme with 150 local stakeholders, including academia, business, community, and government representatives (personal communication, October 2016). The process produced three drafts of the plan, each with a corresponding consultation process. The process produced 22 business plans focusing on different areas of the Council's responsibility, which together formed the final *CPH 2025* document. Relevant units across the Council administration participated in this process and committed to apply the climate goals into their own strategic plans. The final draft was submitted to the Council and adopted in 2012. In keeping with a strategic approach, the plan covers a four-year period that aligns with the budget cycle, so initiatives in the plan are incorporated into the budget expenditure for the four years it will take to implement.

Goal Clarity

The *City of Copenhagen Municipal Plan 2015: A Coherent City* focuses on 'strengthening the quality of life for the Copenhageners and creating growth' (City of Copenhagen 2014b, 1). This links to the overarching goals for the Copenhagen region, which are to maximise economic growth and innovation (Capital Region 2015). The *Municipal Plan*'s objective is to maintain and improve the high quality of life of Copenhagen residents (City of Copenhagen 2014b) and to further enhance the city's environmental performance by making Copenhagen 'the world's first carbon-neutral capital' by 2025 (City of Copenhagen 2012b). These high-level objectives form part of the spatial land use and transport coordination in the *Municipal Plan*. In specific connection to *CPH 2025*, the *Municipal Plan* reiterates the city's commitment to future directions for mobility in the city, including projected distribution of traffic in Copenhagen. Public and bicycle transport are intended to account for at least 30% of traffic each, with cars constituting no more than 30% of all traffic. The Municipal Plan also intends that at least 75% of the growth in traffic must be green and at least 50% of traffic to work or school in Copenhagen be by bicycle.

Measurement

In establishing the original targets and performance measures, the city consulted widely with internal and external stakeholders. The evidence-based approach utilised both scientific knowledge and local understandings to develop business plans across various areas relevant to climate change impacts in the city. The city is a member of both the ICLEI and the C40. Like most members of these networks, the city adopted the UN GHG Protocol to measure emissions. As a leading member of the C40, Copenhagen was a pilot city for establishing the *Global Protocol for Community-Scale Greenhouse Gas Emission Inventories*. The city does undertake some specific measurement practices that reflect the principles of the performance management doctrine. One example is the objective to increase the rate of cycling in the city, which reflects the approach taken by the Council to its measurement methodologies. The approach includes broad community-level consultation, local workshops, and focus groups to propose long-term visions of the city as the 'best' in the world for cycling. The broad goals help to develop the specific projects and initiatives that would be needed to achieve them. *Bicycling Accounts.*

Performance assessments are a reflection of the commitment of the city government to the value of performance measurement. The Council argues that it has established an annual report process with the intention to measure developments and also to relay a common story about the various projects (City of Copenhagen 2014a). According to the Council, *CPH 2025* will result in significant amounts of information and will be accumulating knowledge in numerous fields. In the period up to 2025, there will be three general evaluations of *CPH 2025*, where the two first are intended to provide input for new initiatives and adjustments of the Plan in the subsequent period. The first report was being compiled at the time of writing, as a conclusion of the first implementation period 2013–2016. The period 2017–2020 will be evaluated in 2019–2020. During the period 2025–2026, the final evaluation will be done through an assessment of the goals achieved.

Mainstreaming and Incorporation

The themes of Copenhagen's approach to mitigation highlight the linkages in the other planning documents, suggesting some form of mainstreaming of climate and sustainability considerations across Council functions. As a result of the original internal consultation process, there is a commitment to collaboration throughout the policy implementation process. A 2015 study recognises the effectiveness of the approach being taken where the integrated transport and urban planning organisation in Copenhagen promotes cycling more than the less integrated planning arrangements in other Scandinavian cities (Koglin 2015). Council officers claim, most importantly for the purposes of performance assessment, that

> the themes are also covered in a reporting regime that provides information on progress against the original objectives of CPH 2025. (Personal communication, October 2016)

The regular assessment and evaluation of *CPH 2025* provides the community with the strongest publicly observable evidence that some form of performance management framework is present in the implementation of climate initiatives. Performance reporting is undertaken biannually and is available primarily through a series of *Green Accounts* reports, produced in 2012 and 2014. Additional reports, both general and specific, have appeared, for example *Copenhagen Climate Projects: Annual Report for 2014 & 2015* and *Copenhagen European Green Capital 2014: A Review*. According to senior managers, these reports

> have a specific purpose of trying to communicate directly with the business community in the city that climate initiatives can result in direct economic benefit to business activities. (Personal communication October 2016)

Use of Performance Information

The Council has mixed results in its use of performance information. Within the Council, administration officers claim that performance information is used to promote policy and management change. The evaluation process of *CPH 2025* has successfully utilised performance information to establish new areas where emissions reductions can be achieved. Performance information has been used internally to support greater investment in specific climate initiatives. Despite overall budget commitments, there is still a need for individual projects to be supported by the Council. As an example, the Bicycle Secretariat often needs to use performance data from the *Bicycle Accounts* database to argue for stronger commitments. In pitching for greater resources, the Secretariat needs to develop arguments for investment in cycling in socio-economic terms, with:

> Cost benefit analysis and return on investment calculations forming the basis of submission to council for funding on an annual basis. (Personal communication, October 2016)

Despite the highly complex nature of its arguments, the Secretariat often faces ongoing challenges in succeeding with funding allocations when compared with automobile infrastructure and comparatively high levels of investment in public transport (Freudendal-Pedersen 2015; 602). Senior managers within the Secretariat express some frustration about the need to

> [j]ustify continued investment in bicycle infrastructure when the city government continues to spend ten times more of its budget on automobile infrastructure, and four times more in public transport infrastructure. (Personal communication, October 2016)

The Council has been willing to experiment with various forms of reporting in order to try to raise attention within the building owners' community to changes needed to reduce emissions from buildings. These continue to be advisory and voluntary approaches, and the city government has been ineffective in using performance information to justify the introduction of regulatory measures to reduce emissions from existing buildings. According to one senior planner:

> This has been one area where we have not been as successful as we hoped. (Personal communication, October 2016)

Similarly, the Council has been challenged by the need to raise greater awareness in the community to the work on responding to climate change. One senior manager explained:

> Council research shows only 15–20% of the local population are aware of the CPH 2025 climate projects and that Copenhagen's target is to be carbon neutral by 2025. The council recognises there is considerable work to be done in this regard and is examining ways to improve its communications with the community. (Personal communications, October 2016)

The cycling objectives reflect the city government's communication approach, including broad community-level consultation, local workshops, and focus groups to propose long-term visions of the city as the 'best' in the world for cycling. The broad goals help to develop the specific projects and initiatives that would be needed to achieve them. From the initiatives, the Council established the budget commitments necessary for implementation.

The organisation structure establishes a city administration with related functions falling under leadership of a single Mayor. In the case of climate change, the structure lends itself to the mainstreaming of functions relevant to policy coordination; for example, transport infrastructure and cycling are within the TEC, as are planning and building regulations. Despite this vertical integration and coordination, it is still possible for budget restrictions to prevent coordination of road maintenance to miss opportunities for cycle path development in areas identified for improvements. This level of mainstreaming also supports a degree of incorporation of performance into the management processes, as individual areas operate under the same management and performance objectives. Reporting against targets also becomes more streamlined as the management is accountable for achieving the same targets. There continues to be some challenges however with horizontal coordination across the municipal administration and policy areas. This is a common problem in municipalities in Denmark and is well documented (OECD 2009). Most Danish councils face this challenge and Copenhagen is no exception. However, the fact that the TEC mayor also sits on the Finance Committee helps to support the financial commitment and coordination of the climate initiatives across the functions.

Conclusion

The policy approach in Copenhagen towards climate change continues to be advisory and voluntary, as the Council has not been willing or is unable to justify the introduction of regulatory measures to reduce emissions. This impacts on the application of performance management in the implementation stages in various ways. The effectiveness of coupling of economic and sustainability issues in Copenhagen has been important for the pursuit of a climate policy agenda. An important lesson for other city governments however is that in Copenhagen, this coupling can be a double-edged sword for climate policy and the application of performance management. From the evidence presented here we can see there have been positive results from political commitments to climate objectives that have economic benefits. Such commitments set up an immediate dilemma, and potential conflict, for goal clarity, which is the foundation for a performance-based approach to climate policy. Council managers face a challenging situation if economic objectives override environmental considerations. The links between the green economy and the climate objectives have been politically important for the Council in appearing to promote much needed economic opportunities for local residents while also pursuing objectives of reducing emissions and preparing the city against flooding. The objective of establishing a city environment that will attract the 'creative class' has so far worked in favour of those promoting mitigation and adaptation initiatives. However, the question of the precedence of economic competitiveness over climate goals can be a long-term challenge to environmental goals. If the commitment weakens, there could be a corresponding reduction in the desire to measure performance and report results of efforts to reduce emissions and improve the resilience of the city to anticipated extreme weather events. Whether environment issues will take precedence when economic opportunities are not clearly evident remains to be seen. It could be argued that the warning signs are appearing as traffic infrastructure has precedence over investment in cycling.

Another important lesson for other city governments is the commitment in Copenhagen to a structured evaluation process throughout the implementation of *CPH 2025*. Here, the evaluation and review process helps to establish a double-loop learning process, with new ideas for management change as well as for change to policy objectives. The Copenhagen city government has also been willing to experiment with various forms of performance reporting on the achievements of CPH 2015 in order to try to raise attention of the business and the local community to necessary changes for reducing emissions from energy use, travel, and construction activities. The evaluation process, combined with the consideration of new methods, reveals a level of willingness to attempt policy learning during the implementation stages. This commitment is a strong indicator of the seriousness with which the Council has been undertaking a response to climate change. The senior managers interviewed remain committed to the climate goals and take some pride in their application of performance to the results achieved so far in reducing emissions in the city. It remains to be seen whether this success will provide the foundation for further environment reforms needed to achieve climate goals.

References

Andersen, J., & Pløger, J. (2007). The Dualism of Urban Governance in Denmark. *European Planning Studies, 15*(10), 1349–1367. DOI 10.1080/09654310701550827.

Bayliss, D. (2007). The Rise of the Creative City: Culture and Creativity in Copenhagen. *European Planning Studies, 15*(7), 889–903. DOI 10.1080/09654310701356183.

Berman, B. (2013). Better Place Pulled Down Israel and Denmark Electric Car Markets, Say Local Drivers. *Plugin Cars*. http://www.plugincars.com/better-place-pulled-down-israel-and-denmark-electric-car-market-say-local-drivers-129094.html

Blom-Hansen, J. (2012). Local Government in Denmark and the 2007 Municipal Reform. In A. Moisio (Ed.), *Rethinking Local Government: Essays on Municipal Reform* (Vol. 61). Helsinki: VATT Julkaisu.

Braw, E. (2013). Tackling Climate Change: Copenhagen's Sustainable City Design, *The Guardian*. https://www.theguardian.com/sustainablebusiness/tackling-climate-change-copenhagen-sustainable-city-design

Capital Region of Denmark. (2012). *Climate Strategy for the Capital Region*. https://www.regionh.dk/english/environment/Documents/16582_KLIMA_pixi_2012_engelsk.pdf

Capital Region of Denmark. (2015). *Greater Copenhagen: Regional Growth and Development Strategy*. https://www.regionh.dk/english/businesses/china/background%20information/Documents/Greater%20Copenhagen%20%20Regional%20growth%20and%20development%20strategy.pdf

Cathcart-Keays, A. (2016). Where Is the Most Cycle-Friendly City in the World? *The Guardian*, January 5. https://www.theguardian.com/cities/2016/jan/05/where-world-most-cycle-friendly-city-amsterdam-copenhagen

City of Copenhagen. (2011). *Copenhagen Climate Adaptation Plan*. en.klimatilpasning.dk/media/568851/copenhagen_adaption_plan.pdf

City of Copenhagen. (2012a). *Copenhagen Green Accounts 2012*. kk.sites.itera.dk/apps/kk_pub2/pdf/1052_ZCazXZtCVK.pdf

City of Copenhagen. (2012b). *CPH 2025: A Green, Smart and Carbon Neutral City*. kk.sites.itera.dk/apps/kk_pub2/pdf/983_jkP0ekKMyD.pdf

City of Copenhagen. (2014). Copenhagen City of Cyclists: The Bicycle Account 2014. http://www.cycling-embassy.dk/wp-content/uploads/2015/05/Copenhagens-Biycle-Account-2014.pdf

City of Copenhagen. (2014a). *Copenhagen Climate Projects: Annual Report 2014*. kk.sites.itera.dk/apps/kk_pub2/pdf/1437_jKjlNNQ38N.pdf

City of Copenhagen. (2014b). *The Coherent City: Municipal Planning Strategy 2014 for the City of Copenhagen*. https://www.kk.dk/sites/default/files/uploaded-files/Municipal%20Planning%20Strategy%202014%20-%20english%20version.pdf

City of Copenhagen. (2014c). *Copenhagen City Government 2014–2017*. https://www.petrakellystiftung.de/fileadmin/user_upload/newsartikel/PDF_Dokus/City-Government_COP.pdf

City of Copenhagen. (2014d). *Copenhagen Green Accounts 2014*. kk.sites.itera.dk/apps/kk_pub2/pdf/1393_bpsA36PvpI.pdf

City of Copenhagen. (2015a). *Copenhagen Climate Projects: Annual Report 2015*. kk.sites.itera.dk/apps/kk_pub2/pdf/1327_khT72sMrcI.pdf

City of Copenhagen. (2015b). *Copenhagen European Green Capital 2014: A Review*. http://ec.europa.eu/environment/europeangreencapital/wp-content/uploads/2013/02/Copenhagen-Post-Assessment-Report-2014-EN.pdf

DMCEB (Danish Ministry of Climate, Energy and Building). (2013). *Denmark's Sixth National Communication and First Biennial Report*. Under UNFCCC. https://unfccc.int/files/national_reports/annex_i_natcom/submitted_natcom/application/pdf/nc6andbr1-dnk-2jan2013[1].pdf

EC (European Commission). (2014). *Copenhagen: European Capital 2014*. http://ec.europa.eu/environment/europeangreencapital/winning-cities/2014-copenhagen/

Florida, R. (2008). *The Rise of the Creative Class*. New York: Basic Books.

Freudendal-Pedersen, M. (2015). Whose Commons Are Mobilities Spaces?–The Case of Copenhagen's Cyclists. *ACME: An International E-Journal for Critical Geographies, 14*(2), 598–621.

Government of Denmark. (2013). *The Danish Climate Policy Plan: Towards a Low Carbon Society*. https://stateofgreen.com/files/download/386

Government of Denmark. (2016). *Denmark's Convergence Programme 2016*. http://ec.europa.eu/europe2020/pdf/csr2016/cp2016_denmark_en.pdf

Govlink. (2016). Germany to Ban Petrol and Diesel Cars by 2030 to Reduce Emissions. http://www.australiangovlink.com.au/index.php?option=com_k2&view=item&id=1245:germany-to-ban-petrol-and-diesel-cars-by-2030-to-reduce-emissions&Itemid=239

Green, B., Lynch, J., McGillicuddy, L., Samuel, V., McCauley, S., Taylor, S., & Kinicki, R. (2014). The Danish Consumer Interests Regarding Electric Vehicles. Worcester Polytechnic Institute. https://web.cs.wpi.edu/~rek/Projects/ECar_PQP.pdf

Hansen, A. L. (2009). Space Wars and Urban Governance: Inhumane Creative Urban Transformations. In F. Eckardt & I. Elander (Eds.), *Urban Governance in Europe* (Vol. 2, p. 111). New York: BWV Verlag.

KL (Local Government Denmark). (2015). Analysis: The relation between EU and the Danish municipalities. http://www.kl.dk/ImageVaultFiles/id_72760/cf_202/EU_influence_on_Danish_municipalities.PDF

Koglin, T. (2015). Organisation Does Matter—planning for cycling in Stockholm and Copenhagen. *Transport Policy, 39*, 55–62.

Lai, K. (2017). How Much Warmer Was Your City in 2016? *New York Times*. https://www.nytimes.com/interactive/2017/01/18/world/how-much-warmer-was-your-city-in-2016.html

Levering, P. (2014). Danish Government Unveils Plan to Help Economy Exit Crisis. *Bloomberg News*. https://www.bloomberg.com/news/articles/2014-05-07/denmark-set-to-unveil-growth-plan-to-drag-economy-out-of-crisis

Levering, P. (2015a). Teslas Hit by 180% Danish Tax on Cars as Green Goals Ditched. https://www.bloomberg.com/news/articles/2015-09-29/teslas-hit-by-180-tax-in-denmark-as-green-goals-get-left-behind

Levering, P. (2015b). Denmark's Government Readies U Turn on Ambitious Climate Targets. https://www.bloomberg.com/news/articles/2015-09-01/denmark-s-government-readies-u-turn-on-ambitious-climate-targets

LGDK (Local Government Denmark). (2009). *The Danish Local Government System*. http://www.kl.dk/ImageVaultFiles/id_38221/cf_202/Background_Paper_-_Local_Government_in_Denmark.PDF

LSE (London School of Economics). (2014). *Copenhagen: Green Economy Leader Report*. http://www.kk.dk/da/om-kommunen/indsatsomraader-ogpolitikker/natur-miljoe-og-affald/klima/co2-neutralhovedstad

Lund Hansen, A., Thor Andersen, H., & Clark, E. (2001). Creative Copenhagen: Globalization, Urban Governance and Social Change. *European Planning Studies, 9*(7), 851–869. DOI 10.1080/09654310120079805.

Monocle. (2014). Most Liveable City: Copenhagen. https://monocle.com/film/affairs/most-liveable-city-copenhagen/

Nielsen, T. A. S., Skov-Petersen, H., & Agervig Carstensen, T. (2013). Urban Planning Practices for Bikeable Cities—The Case of Copenhagen. *Urban Research & Practice, 6*(1), 110–115.

Nordregio. (2016). *State of the Nordic Region 2016.* Chapter 6: Unemployment: Young People Pay the Price for an Incomplete Recovery. http://www.nordregio.se/en/Metameny/Nordregio-News/2016/State-of-the-Nordic-region-2016/

OECD. (2009). *Territorial Reviews: Copenhagen, Denmark.* http://www.oecd.org/gov/oecdterritorialreviewscopenhagendenmark.htm

Pearson, C. (2007). Copenhagen. In C. A. Pearson (Ed.), *Conceptions of the Desirable: What Cities Ought to Know About the Future* (pp. 51–57). Vienna: Springer-Verlag.

Ryan, C., & Shankleman, J. (2017). UK Joins France, Says Goodbye to Fossil Fuel Cars by 2040. Bloomberg. https://www.bloomberg.com/news/articles/2017-07-25/u-k-to-ban-diesel-and-petrol-cars-from-2040-daily-telegraph

Staufenberg, J. (2016). Norway to 'Completely Ban Petrol Powered Cars by 2025. http://www.independent.co.uk/environment/climate-change/norway-to-ban-the-sale-of-all-fossil-fuel-based-cars-by-2025-and-replace-with-electric-vehicles-a7065616.html

Sustainia. (2011). *Copenhagen—Beyond Green: The Socio-economic Benefits of Being a Green City.* http://www.sustainia.me/resources/publications/mm/CPH%20Beyond%20Green.pdf

World Bank. (2016). *Doing Business: Measuring Business Regulations: Denmark.* http://www.doingbusiness.org/data/exploreeconomies/denmark

7

Tokyo Municipal Government

Introduction

In 2010, the Tokyo municipal government (TMG) established the world's first mandatory city-level emissions trading scheme. The scheme represents a dramatic shift in policy approach after two decades of attempts by the TMG to encourage the private sector to reduce emissions by voluntary means. The TMG Cap and Trade Programme (TCTP) is part of a comprehensive set of initiatives incorporated in the 10-year *Tokyo Climate Change Strategy* (TCCS) established in 2007. The TCCS established a series of mitigation strategies that focused on reductions in CO_2 emissions in private enterprises, households, urban development, and vehicle traffic (TMG 2007). Most local governments in Japan approach climate change mitigation through soft policy options, such as disseminating information and educating citizens about energy efficiency. The TCCS represents a more serious approach to policy action by using local ordinances to reduce GHG emissions. In the context of discussions about the application of performance measurement in the implementation of climate policy, the TMG has impressed many researchers with its monitoring systems that check the effectiveness of its climate strategies and initiatives (Roppongi et al. 2016; Niederhafner 2013; Rudolph and Kawakatsu 2012). The results show that

S. Jones, *Cities Responding to Climate Change*,
DOI 10.1007/978-3-319-64810-1_7

the TMG climate policies have achieved the 25% GHG reduction target by 2015, five years earlier than the original 2020 deadline (TMG 2016a). Some argue that the approaches taken by the TMG should be 'seriously considered by other major cities in the world in order to foster a sustainable bottom up approach to global climate protection' (Rudolph and Kawakatsu 2012). There can be no denying the significance of the challenge faced by the TMG. Tokyo City, with its 13 million people, is within the 'mega-city' region of Greater Tokyo, with a population of approximately 38 million. The city produces an average of 65 million tonnes of GHG annually, comparable to the emissions produced by Denmark and Sweden. In this context, the achievements of the TMG in reducing emissions warrant closer analysis.

The availability of data, along with mandatory reporting, has been instrumental components of the processes established in Tokyo to achieve emissions reductions targets. The purposeful development of 'in-house' management capacity by the TMG, with a focus on the specialised skills in monitoring, measurement, and negotiation, helped to establish a team capable of working with stakeholders throughout the policy process, from development to implementation (Roppongi et al. 2016; Niederhafner 2013; Rudolf and Kawakatsu 2012). The presence of this team within the Bureau of Environment (the Bureau) has been critical to the ability of the TMG to use data in two critical stages of the policy process; first, to build convincing arguments in the consultation stages, explaining the value of the TCTP for stakeholders, such as major corporations and industry groups (Nishida and Hua 2011); and second, through the production of regular reports that provide feedback and technical details for evaluating and strengthening the design of policies where needed (Niederhafner 2013). Yet, despite what seems like a highly successful approach, the policy initiatives taken in Tokyo have not been widely adopted by other city governments in Japan. For example, only one neighbouring municipality, Saitama Prefecture, has agreed to participate with Tokyo through a bilateral agreement in its TCTP, with credit reductions for small- and medium-sized facilities across both schemes (EDF 2015). In addition, notwithstanding strong lobbying over the last decade, the TMG has not been able to convince the national government of Japan to establish regulations to facilitate similar city-based emissions trading schemes (ETS) across the country

(Roppongi et al. 2016, 13). As a result, despite success in Tokyo, the overall emissions for Japan continue to increase, particularly in the energy sector, despite the presence of national climate policies and a target to reduce emissions by 3.8% below 2005 levels by 2020 (UNFCC 2015).

The characteristics of the situation in Japan suggest that the examination of the TMG will be informative in the context of the theoretical view that local governments can be a laboratory of policy development for climate schemes and initiatives to be tested and for other levels of government to learn from its successes and failures. It is also interesting from a related but more practical view presented by networks such as the C40 and the ICLEI of the leadership role of city governments for effective climate policy. For the purposes of this book, Tokyo is most interesting in its use of performance information for management and policy change, and the apparent success of its mitigation policies in reducing emissions. A determination to be successful in implementing climate policy has also been part of a wider agenda to help improve public trust in politicians and government bureaucrats following a series of scandals in Japan. As we have seen in Chap. 3, trust in government decision-making about climate change is an issue of critical importance to encouraging behaviour change by both business and citizens. Extensive consultation with local stakeholders has been the foundation on which the TMG has developed its emissions reductions strategies. The use of performance data has been a key element in policy development that has seen the TMG establish a successful cap-and-trade scheme for the city.

The approach taken by the TMG's Bureau of Environment (the Bureau) has enabled the establishment of policies that most city governments have been unwilling or unable to establish in the face of powerful interest groups and highly contested climate issues. The establishment of the cap-and-trade system in Tokyo is the result of a long process of engagement with climate-relevant policies by the TMG. To understand the implementation challenges of climate policy faced by the TMG, and the ultimate decision to adopt the cap-and-trade system, the first section of this chapter will examine contextual issues such as the anticipated impacts of climate change on Tokyo City. The second section will examine the roles and responsibilities and current trends in local governments in Japan that have some impact on climate policy. More specific issues in terms of performance management will be presented from the current research, supported by the perspectives of senior Bureau managers, interviewed in November 2016 and June 2017, in the third and fourth sections by examining the approach to strategic planning, policy development, and implementation. This examination will provide insights into the role of the Bureau within the city administration and the application of performance in its operations.

Climate Change Impacts in Tokyo

The TMG *Climate Change Strategy 2007* is rather vague in outlining the impact of climate change on the city. There is some mention of broader consequences of global warming such as 'the frequent occurrence of unusual weather phenomena, difficulty of food production, depletion of drinking water, and the loss of inhabitable land due to the rising sea' (TMG 2007, 2). There is also recognition that a global climate crisis inevitably 'poses a direct threat to Tokyo's very existence. As it has large seaside and coastal areas, the Metropolis may be more vulnerable to the effects of global warming such as a rise in sea levels' (ibid.). This limited assessment reflects the broader picture in Japan, where the first comprehensive report on the impacts of climate change was not completed until 2013 (Ogawa-Onishi and Berry 2013; Hooper 2013). A lack of details on the impacts of climate change contributes to a situation where the TMG has no specific adaptation plans or policies. In fact, observers claim that Japanese local governments have passed few municipal laws and administrative plans to address adaptation, and no practical adaptation measures have been carried out (Tanaka et al. 2011; Hijioka et al. 2016)

Researchers have provided some evidence of the anticipated impact of climate change on Japan and Tokyo. While their observations are an improvement on the situation in 2007, more work is necessary given the significance of Tokyo's economic and environmental impact for Japan. According to these recent assessments, there has been a mean temperature increase of 3.2°C over the past 100 years (Kusaka et al. 2016, 428). This warming effect in the metropolis is at the top of global-scale warming. With ongoing urbanisation and global warming, summer temperatures in Japan have become so severe that the number of heat stroke causalities count more than 1000 per year, exceeding the number of deaths from natural disasters, including typhoons and tornados (Fujibe 2011). In terms of ecological impacts, Japan's Ministry of Environment (2011) reports that more than 30% of amphibians, reptiles, and freshwater and marine species, and more than 20% of mammals and plants are threatened with extinction (i.e., categorised as 'critically endangered', 'endangered', or 'vulnerable'). Figure 7.1 provides a summary of the research, compiled by the World Wildlife Fund (WWF), highlighting many of the anticipated impacts of climate change across Japan.

More specifically to Tokyo, it is the increasing temperatures across the Tokyo Metropolitan Area (TMA) and the effects of heat stress on inhabitants of the city because of urban heat islands that are of greatest concern to policy-makers (Adachi et al. 2012). Urban warming is conspicuous in large

Projected Climate Change	**Impact to humans**
Temperature increase of 2 to 3°C over the next 100 years for all of Japan • Temperature increase of 4°C around the Sea of Okhotsk over the next 100 years • Increase in the number of extreme hot days (days with temperatures exceeding 35°C) • Decrease in the number of frost days by 20 to 45 days per year • Increase of mean precipitation by more than 10% over the 21st Century • Increase in summer precipitation by 17 to 19% • Increase in heavy precipitation events in Hokkaido • Increase in the frequency and intensity of extreme weather events, such as tropical cyclones, heat waves, and heavy rainfall events	Increase in heat wave intensity and heat stress, putting vulnerable populations, such as the aged, at risk • Increased likelihood of infectious and vector and water-borne diseases • Expansion of dengue fever into Hokkaido • Increased allergies and allergy-related diseases • Increased cost of living and protection from more extreme weather events • 67 to 70% increase in wind-related losses from more intense typhoons • Deteriorated freshwater systems and increases in chemical nutrients affecting fish production and harvests from warming temperatures and changes in precipitation • 1.2 to 3.2% increase in the demand for water supply (with a 3°C warming) Negative impacts to fruit crops and an increase in abnormal fruits • Potential temporary increased yield in grain harvests in Hokkaido • 40% decrease in rice yields in central and southern Japan • Potential northern shift of some fish species and changes in the fish abundance and diversity • Declines in snow cover and sea-ice extent will negatively impact winter-dependent tourism

Fig. 7.1 Predicted impact of climate change on Japan
Source: Adapted from WWF (2008)

cities in Japan, reflecting their rapid growth during the last century. The temperature increase in Tokyo, where the meteorological observatory is located in the central business area of the city, has been about three degrees over the last century. Recent numerical studies have revealed widespread urban warming around Tokyo. This phenomenon, sometimes called the extended heat island, is explained by the enhanced surface heating over a large urban area, as well as a reduction in sea breeze penetration caused by increased surface convergence (Fujibe 2011). Mitigation of the urban heat island should be one of the effective ways of adapting to the future warming caused by global climate change. Improved land use planning around cities to, for example, create more green space could be one of the ways to adapt to global warming by suppressing urban heat islands. Japan is in a particularly precarious situation in terms of electricity supply with the shutdown of many nuclear power plants and rising summer temperatures have exacerbated the problems (Kusaka et al. 2016). For every one-degree increase in temperature, electricity demand elevates by 1,900,000 kWh (kilowatt-hour) (Goto et al. 2004), which is equivalent to the energy produced by two nuclear power plants. Given these serious impacts, there is a great demand for accurate and reliable climate projections concerning the summertime thermal environment, especially August, in the TMA. Japan's power supply is predominantly (90%) from fossil fuel–based power stations, oil, coal, and liquefied natural gas:

> CO_2 emissions from power generation have increased by about 65% since FY1990 (290Mt-CO2/yr), mainly because of increased electricity demand (880TWh in FY2012 compared to 748TWh in FY1990) and the increased use of fossil fuel-fired power plants following the Fukushima nuclear disaster (0.53kg-CO2/kWh in FY2012 compared to 0.39kg-CO2/kWh in FY1990). The increase in electricity use was observed mainly in the residential and commercial sectors, which together are responsible for more than 60% of power sector emissions. (Kuramochi 2015, 1327)

Increasing temperatures are highlighting concerns that without a stronger commitment to reducing emissions by increasing the proportion of electricity from renewable sources (3.5% in 1990 to 4.6% in 2012), national climate initiatives will be unable to reduce GHG emissions in accordance with Japan's international commitments (IEA 2008; Takai 2014). Despite the presence of a feed in tariff to promote renewable sources, primarily solar, the comparatively high cost and lengthy approvals processes, up to four years, have discouraged faster development (Kimura and Zissler 2016; Kuramochi 2015). Japan's original emissions reduction target under the Kyoto Protocol was a 6% reduction on 1990 levels by 2020. In 2014, as a recognition that the measures introduced would not be sufficient to meet the target, this was reviewed and downgraded to 3.8% below 2005 levels by 2020 (Takai 2014; Wang-Helmreich 2015).

It has been the failure of national government policies that has driven many local governments in Japan to pursue their own climate policy agenda. There are mixed opportunities and constraints within the Japanese system that provide potential for mitigation and adaptation action by municipalities. There is however a complex relationship between the national government and municipalities in the context of climate policies—these will be examined in the following sections.

Role, Responsibilities, and Trends in Local Governments

Chapter 9 of the *Constitution of Japan* clearly defines the legal basis for local governments. Article 94 outlines the right of local governments to manage their 'property, affairs and administration and to enact their own regulations within the law'. The *Local Government Law* (LGL) is the core national legislation that reflects the Constitutional responsibilities of the local government regarding residents' affairs, the election of councils, the appointment of executive bodies, the status of the local government and its relationship with the national government, and the legal provision for financial arrangements. In Japan, the local level of government is split between regional prefectures and local municipalities. Both operate as presidential systems, where prefectural governors and municipal mayors operate as elected presidents within their jurisdictions (Horiuchi et al. 2015). Organisational structures and intergovernmental relations between municipalities are similar across Japan, but there is considerable difference in the size of prefectures, with Tokyo being the largest, with a population of approximately 13 million, and Tattori the smallest, with

600,000. There are fiscal transfer arrangements in place to help ensure a level of parity in the services provided by both large and small local governments. Service delivery arrangements require joint approaches in many areas largely due to the revenue-raising capacities of the national and local governments. Despite only raising two fifths of government revenue, the range of local government functions is extremely broad, covering all aspects of the country's domestic life other than national security, legal affairs, and international relations (CLAIR 2016). As an example of the range of services provided by large prefectures, Fig. 7.2 outlines the organisational structure of the TMG and the range of responsibilities covered by the municipal government.

Local governments in Japan have been subject to a number of reforms since the 1990s. The national government pursued an amalgamation agenda driven largely by the population movement—from the countryside to the city, an aging population, population decline, and fiscal income decline. Small municipalities without stable incomes became costly for the national government to maintain (Shimizu 2012). A process of merging smaller municipalities produced a decline from 3252 municipalities in January 2000 to 1847 in April 2006. Also 1992 cities, towns, and villages were consolidated into 587 municipalities in the same period. As a result of these mergers, there has been a drop of 30% in the number of municipal politicians across the country (Horiuchi et al. 2015). These changes, like similar amalgamation processes in other countries, have transformed the quality of local autonomy in Japan. Japanese researchers argue that the concept of local government has been changed by a series of amalgamations such that it has become merely a service provider, whose efficiency is very important, while democracy has been weakened (Okamoto 2012). Surveys of local mayors in 2015 reveal contrasting results, with 66% declaring that their expenditures have been reduced, 33% suggesting that their policy-making had improved, 27% believing that service standards had worsened, and 13% arguing that their expenses had increased (Kudo 2015).

The decentralisation objectives of the national government were put into force through the *Omnibus Decentralisation Act of 1999*. Of particular relevance to the discussion of climate change policy is the provision within the Act for local governments in Japan to adopt their own environment

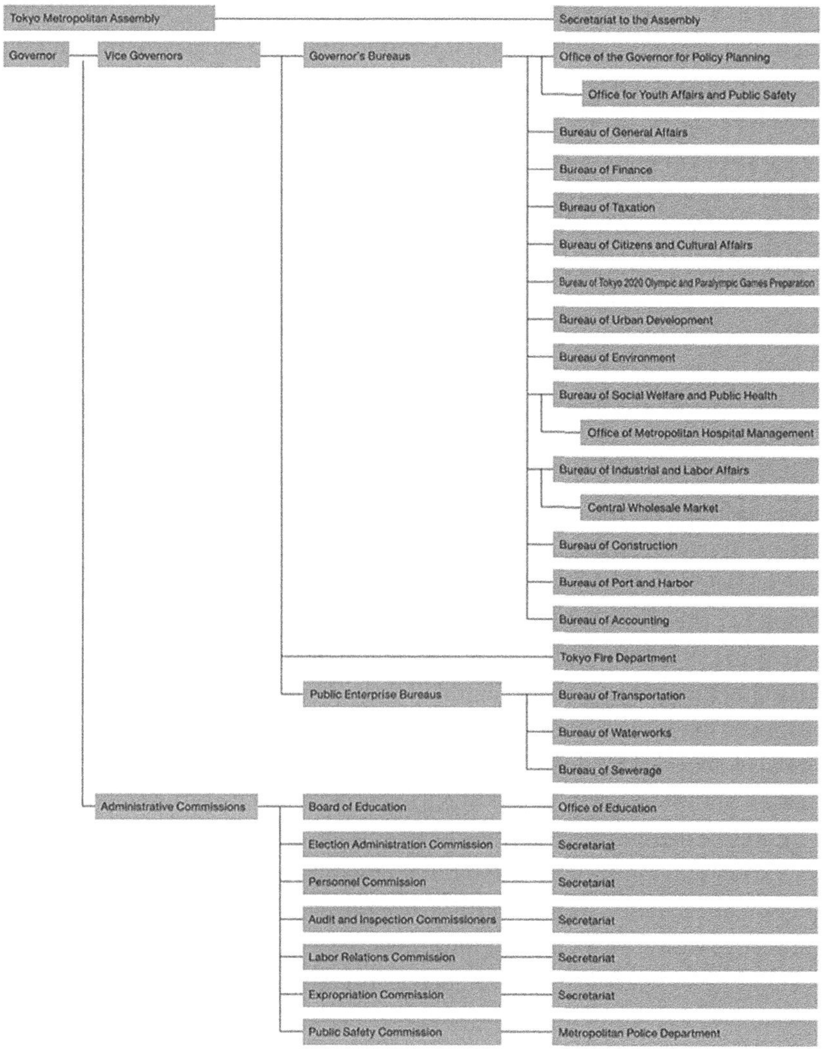

Fig. 7.2 Structure of the Tokyo municipal government (2016b)

policies. Various municipalities have pursued their environmental objectives by introducing taxes in areas such a forestry and waste management (Rudolph and Kawakatsu 2012). Through the Act, local governments have the authority to take legislative action when the 'national government itself has not enacted any specific policies and measures towards climate change, and the national government does not prohibit them from doing so' (Sujiyama and Takeuchi 2008, 429). The TMG has been a leading local government in using the Act to establish mandates for businesses and industries to reduce CO_2 emissions, and to introduce its emissions trading scheme. Climate policy in Tokyo emerged alongside the recognition that the prominent national climate legislations in Japan, known as the Energy Conservation Law and the Global Warming Law, were too weak to bring significant emission reductions (Kuramochi 2015). The opportunity and the capacity of the TMG to undertake climate change initiatives have largely emerged from the action and inaction of the national government.

Executive authority in the TMG rests with the publicly elected Governor, who represents the metropolis of Tokyo. With a four-year term of office, he or she has overall control of metropolitan affairs, and the authority and responsibility for managing the metropolitan administration (TMG 2017). The Governor is assisted by Vice Governors, who in turn have responsibilities in key functional areas and for specific government bureaus (departments). The Tokyo Metropolitan Assembly, consisting of 127 elected members, is the formal decision-making body for the city. The Assembly has the authority to enact, amend, and repeal metropolitan ordinances, approve the budget, and certify its settlement (TMG 2017). The Bureau has overall responsibility for the climate change strategy, including the evaluation and production of public progress reports. The Bureau consists of a number of divisions responsible for areas such as climate change and energy, environmental improvement, natural environment, and sustainable materials management. The Bureau is responsible for the development and implementation of policy commitments, including initiatives to address climate change, electricity and energy issues, motor vehicle pollution, air, water and soil contamination, chemical substances, and wastes and the assessment of environmental impact within the Tokyo city boundaries.

In the early 1990s, Japanese municipalities began searching for ways to improve the level of trust in both politicians and bureaucrats. A series of scandals at both the national and the local level of government had seen trust in the government fall to historic lows due to a failure to create policy to adequately respond to the global economy, the needs of civil society, and scandals involving bribes (Kikuchi 2007; Kim 2010). Many local governments introduced NPM-based initiatives, including performance measurement, programme evaluation, customer satisfaction surveys, outsourcing and contracting out, and the general revision of service delivery methods (Kudo 2015). The national government was also looking for ways to reduce its costs by transferring functions to local governments; municipalities were forced to provide services directly financed by their citizens. Performance measurement in municipalities has served two key functions of, first, providing evidence to citizens of good performance and service and, second, of providing information for policy and programme evaluation. There have been increasing efforts to promote transparency through stakeholder consultations that enable citizen advocacy (Tsujiyama 2002). As we will see later in the chapter, these elements of NPM have been critical to the introduction of the cap-and-trade scheme by the TMG.

Survey data from over 1000 mayors across Japan reveals that the current and future priorities for local governments include establishing a sound financial position for their municipality (28%), revitalising the local economy (21.3%), and countermeasures for a declining population (17.9%). Most fear that in 10 years, their financial situation would be worse (75.9%), due to a change in subsidies (37.8%), an increase in social welfare–related expenditure (23.1%), and a decrease in local tax revenue (21.5%) (Kudo 2015, 158). Inherent within the NPM reforms has been the introduction of strategic planning as a method of improving performance and demonstrating the effectiveness of policy initiatives. The development and implementation of climate policy in the TMG has been based on a strategic planning approach that includes considerable stakeholder consultation, the drafting of action plans, regular reviews and evaluations, and a series of public reports on the progress of policy objectives. Climate change has been a contentious area of public policy, with considerable differences between national- and local-level approaches. The TMG has been at the forefront of pushing the climate agenda and, in many instances, has been willing to go further than the national government in taking a regulatory approach to reduce emissions in the city (Nishida 2013). Despite these differences, local governments in Japan are reliant on the national government to set the policy boundaries and provide the resources to implement climate initiatives. The next section outlines the national climate plans that provide an institutional context for the climate change initiatives of the TMG.

Climate Policy Action in Japan

In Japan, the national government can choose to delegate authority to governors and mayors to implement many national environmental laws. The approach taken by the government in implementing its GHG reduction targets has been to ask municipalities to develop and implement action plans that reflect their local conditions. There are a number of government ministries with responsibility for climate policy—see Table 7.1; however, the main responsibilities rest with the Ministry of Environment (MOE) and the Ministry of Land Infrastructure and Tourism (MLIT). Despite the reforms of the 1990s, Japan remains a centralised intergovernmental system where local governments are largely responsible for policy implementation (Kanai 2015). This characteristic does not however restrict opportunities for local governments that have a desire to achieve policy goals according to their own objectives. Such policy initiatives will be assessed by the national government according to its policy commitments. There is potential for conflict and power struggles in these circumstances. Where conflict occurs, local governments need to keep national ministers and agencies 'on side' by establishing a reputation for quality policy-making. In the case of climate policy, many local governments in Japan find it important to achieve 'reputation, support and even fame' among their peers within the local government community (ibid.; 153)

Table 7.1 Government ministries with climate change responsibilities in Japan

Ministry	Responsibilities
Prime Minister and the Cabinet	Global warming headquarters Implementation and evaluation of climate change policies and measures
Ministry of the Environment	Climate policy
Ministry of Economy, Trade and Industry	Energy policy
Ministry of Land Infrastructure, Transport, and Tourism	Development of the *Low-Carbon City Act*, energy-efficient houses and buildings, energy generation, storage and savings, urban environmental planning
Ministry of Foreign Affairs	Representation of Japan's position in international climate negotiations

because this establishes a level of credibility with relevant national ministers who can provide support to help improve chances of policy success. This is important for local policies that help achieve national objectives; while meeting local needs, municipalities can receive national subsidies and political support.

The MOE is responsible for the *Law Concerning the Promotion of the Measures to Cope with Global Warming*, within which Article 4 states the responsibilities of local governments in the implementation of climate mitigation policy:

1. Local governments shall promote policies to limit GHG emissions and to enhance sinks in accordance with the natural and social conditions of their areas.
2. Local governments shall take measures to limit GHG emissions and to enhance sinks in their own business activities, strive to provide information on policies specified in paragraph 1, and adopt other measures so that enterprises and residents in their areas are encouraged to take up activities to limit GHG emissions and to enhance sinks.

These are not mandated requirements but expectations by the national government about what local governments in Japan should include in their climate mitigation plans. In this scenario, local governments are expected to establish plans to improve traffic flow, promote energy saving in commercial buildings and the business sector, provide support for measures deployed at the municipal level through their centres for climate change action, and recruit volunteers to promote climate change actions (Sugiyama and Takeuchi 2008, 428). In terms of the measurement and evaluation of the actions of local governments in terms of their own business activities, the MOE has established a performance measurement framework that covers the following limited range of requirements:

1. Prefectures and cities, towns, and villages shall formulate plans on measures to limit GHG emissions from and enhance sinks in their business activities (hereinafter referred to as the action plan in this Article), in line with the basic policy.

2. When formulating or changing these action plans, prefectures and cities, towns, and villages shall publicise them without delay.
3. Prefectures and cities, towns, and villages shall publicise the state of the implementation of the measures specified in their action plans (including the aggregate emissions of GHGs).

The national government established the *Basic Environment Plan*, in accordance with the provision of Article 15 of the *Basic Environmental Law* (Law no. 91, 1993). The most recent *Basic Environment Plan* from 2012 prescribes four broad, long-term objectives: (1) to build a socio-economic system fostering an environmentally sound material cycle, where the environmental load by human activities is minimised; (2) to secure a harmonious coexistence between humankind and diverse wildlife and natural environment; (3) to enhance participation of all members of society in environmental conservation activities; and (4) to enhance their international activities (MOE 2016). The section 'Role, Responsibilities, and Trends in Local Governments' also provides some vague details on the intentions of the government to establish 'comprehensive indicator/indicators which show the progress of these objectives and the relation between the objectives and measures, in order to ensure effective implementation of the measures' (MOE 2012). By 2016, the MOE website claimed that the 'Government will immediately begin working on development of the comprehensive indicators and utilize the results in implementing and reviewing the Plan' (MOE 2016).

The MLIT established a *Low-Carbon City Plan* under the 2012 *Low-Carbon City Promotion Act* (Act no. 84, 2012). These Acts do not prescribe mandatory requirements but expectations that local governments begin to formulate low-carbon city plans. The national approach is to promote low-carbon urban development and more efficient city administration. It is an incentive-based approach, with tax concessions and financial assistance for local governments. The *Low-Carbon City Act* requires municipalities to formulate Low-Carbon City Development Plans that include energy-saving proposals, reductions in infrastructure costs, urban heat island mitigation, and increased resilience against disasters (MLIT 2014). The Act is aimed at residential, commercial, and transport sectors, which, when combined, produce approximately 50% of Japan's CO_2 emissions.

Some cities have been successful in reducing emissions by focusing their climate policies in the areas encouraged by the Act. One example is Omihachiman City, which has supported more compact urban structures and centralised urban functions. Success has been attributable to serious attempts by the local municipality, working with the financial support from the national government (Suzuki et al. 2015) (Table 7.2).

International organisations, such as the WRI and the OECD, maintain that Japan's non-mandatory approach in the climate measures it supports will continue to be insufficient to reduce emissions to meet its own climate objectives. For example, in 2009, the OECD claimed:

Table 7.2 Guiding policies for climate change for the Tokyo Municipal Government (TMG)

National Global Warming Countermeasures Promotion Act	Stipulates that local governments are required to develop their own greenhouse gas (GHG) emissions mitigation plans in accordance with the national mitigation plan (see Section V, Subsection 'Major Subnational Policies'). The national mitigation target is not divided among prefectures (Sugiyama and Takeuchi 2008); instead, local governments decide their mitigation target levels on their own.
National Basic Environment Law and Basic Environment Plan	Promotes systematic and comprehensive policies for environmental conservation. Encourages local governments to build a socio-economic system fostering an environmentally sound material cycle, where the environmental load by human activities is minimised, and to secure a harmonious coexistence between humankind and diverse wildlife and natural environment.
Low-Carbon City Promotion Act 2012	Promotes cross-sectoral emissions reductions through the development of compact cities to achieve both low-carbon urban development and more efficient city administration.
Tokyo Environmental Master Plan 2016	Is an Environmental Master Plan that showcases the environmental policies to be implemented by 2030. With ambitious goals, such as the GHG emissions reduction target and the renewable energy target, the TMG will commit to actions for a sustainable Tokyo at full scale and promote the engagement of citizens, businesses, and other stakeholders to achieve these goals.

(*continued*)

Table 7.2 (continued)

Tokyo Green Building Programme	Proposes that new buildings with 5000 m^2 of total floor area or larger in Tokyo be evaluated on their environmental performance, followed by public disclosure. Includes environmentally sensitive guidelines for the construction and performance ratings for new buildings across the city, and information regarding related programmes such as the Certificate Program for Energy Conservation and the Green Labeling Program for Condominiums.
Diesel Replacement Scheme	Proposes subsidies to help local businesses replace or retrofit non-compliant diesel vehicles introduced in 2001. Enforcement through heavy fines as well as naming and shaming began from late 2003 in Tokyo. Upper limits on diesel exhaust were lowered further in 2006.
Tokyo Climate Change Strategy 2007	Is a 10-year mitigation strategy to reduce Tokyo's GHG emissions by 25% below 2000 levels by 2020. Initiatives focused on reductions in CO_2 emissions in private enterprises, households, urban development, and vehicle traffic.

Emissions are to be limited through comprehensive plans by prefectures and certain major cities and reports from businesses and franchises, *but there is no enforcement mechanism to ensure that the plans are achieved* (emphasis added). In sum, it is important to reverse the upward trend in the emissions of the commercial and residential sectors, which account for about one-third of the total. At the same time, further reductions are needed in the industrial sector, which accounted for 38.6% of emissions in 2007. (Jones and Yoo 2009, 17)

The WRI made a similar observation in 2014, arguing that in Japan,

[l]ocal governments have limited authority to independently formulate climate policies because the GW Countermeasures Promotion Act provides them neither the legal competence nor the support necessary to reduce their GHG emissions. Therefore, the local government actions have been limited to the formulation of action plans and awareness-building activities, and it is only in recent years that some local governments started using local ordinances to commit to more concrete climate actions. (Kuramochi 2014, 23)

Tokyo Municipal Government's Climate Change Strategy

In 2007, the TMG was critical of the national government and argued that the 'Japanese government fails to come up with medium and long-term reduction targets or effective and specific measures' (TMG 2007, 3). The city government argued that this failure at the national level had necessitated a new range of local mitigation initiatives that would help reduce emissions across the city:

> In order to step up measures to cope with climate change, which can afford no further delay, Tokyo Metropolitan Government advances the world's highest level strategies in this 'Tokyo Climate Change Strategy' on behalf of the Japanese government and takes the lead in Japan's climate change mitigation measures. (TMG 2007, 3)

The TMG released an *Environmental Master Plan* for Tokyo in 2008. The clear goal of the Plan was to move '[t]oward the city with the lowest environmental impact in the world' (TMG 2008, 1). The focus of the sustainability initiatives by the TMG would:

1. increase focus on environmental sustainability measures,
2. promote systems and mechanisms ahead of those of the central government, and
3. increase the emphasis on environmental policies that allow private enterprises and Non Profit Organisations (NPO) to play a greater public role.

The development of a *Climate Change Strategy* was a key element of the *Master Plan* and represented a 10-year strategic approach by the city government, led by Governor Shintaro Ishihara. Senior managers in the Bureau argue that the clear goals established by Governor Ishihara from 1999 to 2012 were critical to the successful introduction of a mandatory cap-and-trade approach taken by the city. International commentators such as the *Harvard Business Review* argued that Ishihara demonstrated how 'real leadership can make those feigning to lead on climate change look feeble' (Berinato 2008). The Governor had previously been the head of the national environment agency in the late 1970s when Japan was dealing with the effects of the Minamata mercury poisoning, the worst case of industrial pollution in Japan's history (Lam 2011). In an interview on this issue, Ishihara claimed: 'I saw the kind of tragedy that the government can cause when it fails to act' (FT 2008). In a top-down approach, the Governor decided to impose strict limitations on local GHG emissions. In doing so, he risked clashing with the national administration run by his own conservative political party (FT 2008). One interviewee from the Bureau claimed that the Governor made it clear that he would pursue his climate agenda 'even if we are sued for doing so' (personal communication, November 2016). In arriving at this however, the TMG had already been through a series of voluntary initiatives since the mid-1990s, including the Tokyo Regional Plan for the Promotion of Measures to Prevent Global Warming, the Tokyo Metropolitan Environmental Master Plan, the Stop Global Warming, the Tokyo Strategy, and the revision to the Tokyo Metropolitan Environmental Security Ordinance for strengthening CO_2 emission reductions and heat island effect control:

> Despite these measures, Tokyo's CO_2 levels were about 6% above 1990 levels. Due to changing lifestyles (larger office buildings, more single-person households, and office and home automation), emissions rose by about 33% in the business sector and 16% among households. (Schreurs 2010, 95)

When introducing the TMG *Climate Change Strategy* in 2007, Governor Ishihara announced a target to reduce GHG emissions by 25% of 2000 levels by 2020. The Strategy incorporated those initiatives already introduced since 2000, including the *Green Building Programme* and the *Diesel Reduction Plan*. Early assessments of the results of these initiatives revealed that they would be insufficient to achieve the 25% target. Building owners had been slow to adopt measures that would reduce the high level of emissions from their buildings. In 2010, the TMG, under the direction of Mayor Ishihara, went one significant step further and introduced a mandatory cap-and-trade scheme focused on the highest-emitting buildings in the city. The decision to take this mandatory approach has been critical to the success of the TMG's GHG reduction target. Each of the initiatives has been important in reducing emissions in the city, and thus, each warrants further analysis.

Green Building Programme

In the case of the *Green Building Programme* (the *Programme*) introduced in 2000, new public and private buildings with 5000 m^2 of total floor area or larger in Tokyo needed to be evaluated on their environmental performance and to disclose this information publicly. The intention of the *Programme* was to direct building owners and the construction industry towards building in a more environmentally friendly way to reduce emissions and lower demands on the energy system. Residential and commercial buildings in Tokyo consume over 60% of all energy in the city (Nishida 2013). The goal of the *Programme*, in keeping with the overall *Climate Strategy*, is to reduce emissions from buildings by 25% by 2020 and reduce energy consumption by 20% by 2020. In 2005, the *Programme* was revised and new emphasis was placed on climate change and on reducing the heat island effect. With the support of Governor Ishihara, the TMG adopted a regulatory approach and passed eight laws and ordinances covering thermal loads of buildings, the use of natural energy, energy conservation, extension of building life through maintenance and management, water use, and the establishment of green space. Figure 7.3 highlights the areas covered and the initiatives included in the *Programme*.

Since 2002, more than 1500 buildings have been included in the *Programme*, resulting in improvements in insulation and the use of energy-efficient equipment. The TMG's mandate requires that building owners must report and publicly disclose information regarding energy use and efficiency. The performance and monitoring scheme developed and implemented by the TMG's Bureau of Environment provided measurement details for those buildings that volunteered to participate in the early stage of the scheme. Resources posted online were derived from the information gathered by the Bureau. New guidelines were established for the construction and performance ratings for new buildings across the city. The information also contributed to the development of programmes

Category	Items
Energy	Heat load resistance of the building envelope Renewable energy Energy efficient building system (equipment) Efficient control systems
Resource Material	Eco materials Protection of ozone layer Longer building life expectancy Hydrological cycle
Natural environment	Landscaping and bio-diversity
Heat island effect	Atmospheric heat waste Surface of ground and buildings Wind environment

Fig. 7.3 TMG Green Building Programme
Source: Nishida (2013)

such as the *Certificate Program for Energy Conservation* and the *Green Labeling Program for Condominiums.* Senior Bureau managers interviewed for this research claim that the experience of the *Programme* helped in the preparation for developing a proposal for a cap-and-trade scheme to be directed towards buildings with highest emission rates in the city (personal communication, November 2016). In 2005, officers in the Bureau began a series of discussion papers on a mandatory cap-and-trade scheme for large buildings in the city. The performance information gathered from the experience of the *Green Building Programme* was critical to their arguments. According to senior planners in the Bureau, the idea was 'killed off' after the national chamber of commerce (the Keidanren) put pressure on local politicians to stop it, fearing that the local scheme could be adopted nationally (Fig. 7.4).

The Diesel Replacement Scheme

In 1994, the Japanese government mandated a series of new standards for emissions from new diesel vehicles in terms of their nitrogen dioxide (NO_2) and particulate matter (PM). The national legislation introduced in 2002 prohibited the registration of non-compliant vehicles in metropolitan areas. In 1999, Governor Ishihara established an advisory board to provide him with advice on how to strengthen the national legislation

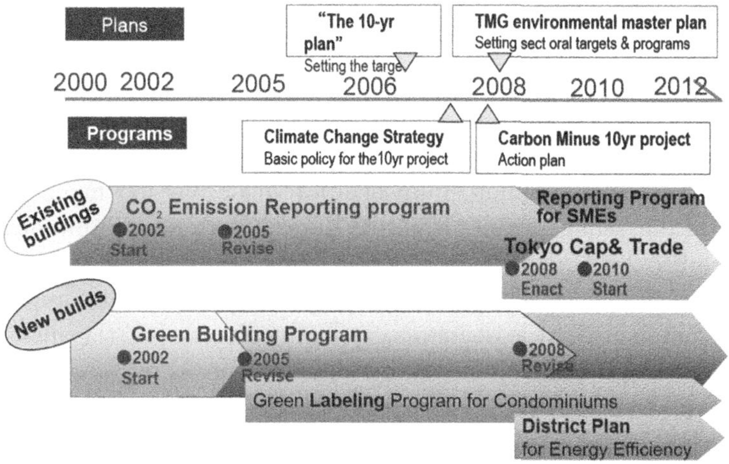

Fig. 7.4 Linking the Green Building Programme and the Climate Strategy
Source: Nishida (2013)

in its application to the use of the existing older diesel vehicles in Tokyo. The advice of the board resulted in the Governor taking a top-down approach to go beyond the national legislation and prohibit vehicles that did not fulfil the emissions standards from driving within the TMG boundaries (Yorifuji et al. 2011). The TMG legislation exhorted citizens (1) not to ride, purchase, and sell a diesel car and (2) to accelerate the development of a device and its installation in diesel cars for cleanup of waste gases and particulates (Hosomi 2010; TMG 2011). Large cities such as Osaka and Nagoya also accepted the diesel exhaust emission control measures instituted in Tokyo. In a broader regional approach, the TMG and neighbouring three prefectures introduced a much stronger ordinance in October 2003 requiring any diesel vehicles travelling in their area to conform to their standards for PM. Senior managers in the Bureau said that they recognised that emission 'leakage' from adjoining municipalities would be an issue that would threaten the success of their efforts. The TMG and Saitama Prefecture further tightened these standards in April 2006 to the level of the 2003 national requirement for new vehicles. Some argue that the top-down decision by Governor Ishihara led to the revision of the national legislation to reduce PM emissions from automobiles in specified areas (Hosomi 2010).

According to TMG regulations, diesel vehicles that did not meet the standards were required to be replaced by new compliant vehicles or required to fit a device to reduce the emission of PM (TMG 2011). Following the introduction of the regulation, all non-compliant vehicles were prohibited from travelling in the Tokyo area. The regional approach adopted by the adjoining municipalities 'made our job much easier', claimed one senior planner from the Bureau (personal communication, November 2016). The TMG regularly inspects diesel vehicles on the roads, as well as at transportation companies, to monitor compliance. Non-compliant vehicle operators can be fined up to JPY (Japanese yen) 500,000. Figure 7.5 shows that the diesel initiatives introduced by the TMG have contributed to a continued reduction in concentrations of suspended particulate matter (SPM) in Tokyo.

Behaviour change initiatives introduced by the TMG included a promotion and public relations campaign as one of its efforts for regulating diesel vehicles. These initiatives included onsite guidance to the approximately 3800 businesses in the area that owned 20 or more vehicles, sending more than a total of 5.5 million pieces of direct mail to diesel car owners in Tokyo, visiting 100 shippers' organisations all over Japan, conducting surveys of 2000 major businesses nationwide on their current regulation measures, and a crackdown on transport delivery points at a total of 300 places. While making efforts to ask diesel car owners and their clients to comply with the ordinance, the TMG tried to raise citizens'

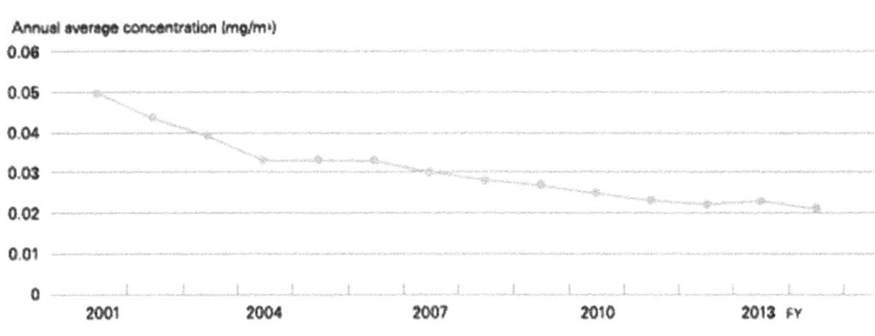

Fig. 7.5 Suspended particulate matter (SPM) since 2001 in Tokyo
Source: TMG (2017)

awareness by conducting an online policy debate, 'Diesel, Yes or No', as well as a public debate (JFS 2011). Small- and medium-sized businesses are eligible for TMG loans and subsidies for installation of PM reduction devices (Takahashi 2011). Violators were detected with monitoring cameras at various points around the city because many of the targeted diesel vehicles, such as trucks and buses, repeatedly take the same routes. Such vehicles were identified through their licence plate car registration and vehicle pollution inspectors delivered administrative directions to the vehicle owners (JFS 2011). Independent assessments of the pollution levels in Tokyo confirm that a 'decrease in air pollutants occurred probably as a result of local government action' (Yorifuji et al. 2011, 3627).

Cap-and-Trade Scheme

The introduction of a mandated cap-and-trade scheme for some of the largest buildings in the city has been widely recognised as the most significant component of the TMG *Climate Strategy* (Rudolph and Kawakatsu 2012; Nishida et al. 2016; Roppongi et al. 2016). The TMG *Environment Master Plan* establishes an overarching target for reducing GHG emissions in the city by 25% below 2000 levels by 2020. To achieve this goal, the Bureau established a three-phase programme that would include the transition from voluntary to mandatory obligations for large-scale commercial and industrial facilities—see Table 7.3. In 2000, a mandatory reporting programme was introduced through the enactment of the *Tokyo Metropolitan Environmental Security Ordinance*. In this stage of the programme, large-scale facilities were required to report their emissions and devise a plan to reduce them by 2020. Reporting was mandatory, but acting on the plans to reduce emissions remained voluntary. The reporting programme provided the TMG with direct measurement of emissions from the largest buildings in the city. In this regard, the TMG was provided with high-quality data (i.e., not derived from proxies). The TMG focused on energy consumption by these buildings, as it accounted for 95% of GHG emissions impacting on the city. This was an efficient and effective approach to take, as

Table 7.3 Progress towards a mandated Tokyo municipal government (TMG) cap-and-trade scheme

Phase 1	
December 1999	Announcement of Tokyo Metropolitan Environment Security Ordinance
April 2000	Implementation of Carbon Reduction Reporting Programme (Phase 1) • Mandatory reporting of emissions and emissions reduction plan • Voluntary emissions reduction
Phase 2	
March 2005	Amendment of Tokyo Metropolitan Environmental Security Ordinance Reinforcement of Tokyo Carbon Reduction Reporting Programme
April 2005	Implementation of Carbon Reduction Reporting Programme (Phase 2) • Introducing a mechanism to provide guidance and advice to the reduction plan • Evaluating the plan and awarding outstanding facilities • Publicising reduction plans (by the TMG and individual facilities)
Phase 3	
May 2007–March 2008	Deliberation at the TMG Environment Council on the introduction of a compulsory cap-and-trade scheme in Tokyo
June 2007	Announcement of the Tokyo Climate Change Strategy • Introduction of the mandatory emissions reduction programme for large facilities proposed
July 2007–January 2008	Stakeholder meetings • Discussions to introduce the Tokyo Cap and Trade Programme
June 2008	Passage of a bill to amend the Tokyo Metropolitan Environmental Security Ordinance
April 2009	Enactment of the amended ordinance and regulations
April 2010	Launch of the mandatory reduction programme • Tokyo Cap and Trade Programme introduced

Source: Adapted from TMG (2012)

[a]pproximately 1300 regulated large business facilities, which only account for 0.2% of the total number of business establishments, are responsible for 40% of the CO_2 emitted from all industrial and commercial facilities in Tokyo. (Roppongi et al. 2016, 5)

The Tokyo Cap and Trade Programme (TCTP) moved into its mandatory phase in 2010 with the establishment of two five-year phases ending in 2019. The scheme focuses on facilities with electricity consumption greater than 15 GWh during a fiscal year. Most electricity (90%) is generated outside Tokyo, so producers are not subject to the scheme. Facilities subject to the scheme in the first phase were required to reduce their emissions by 8% against their 2000 baseline. In the second phase, the target is set at 17% reduction. Measures are taken of energy input and CO_2 output. The figures provided to the TMG by the facilities are verified by officially certified third-party verification agencies. Figure 7.5 outlines the basic elements and operation of the TCTP, including the verification and approval processes. As early as 2013, the results proved that the TCTP had been successful in reducing emissions from the facilities and the participating companies had exceeded their imposed reduction targets of 6–8%, with actual reductions of 23% (Niederhafner 2013).

The trading component of the TCTP was designed as a supplemental measure to be utilised only when the covered facility could not meet its reduction obligations. Under the scheme, the facilities have to meet their reduction goals before they can start selling their credits. The supply will build over the period of the scheme, but only if the facilities reduce their emissions beyond their own target. The offset credits are limited to Japanese credits only, and the TMG is responsible for verifying and accounting for them. The fact that the majority of facilities have exceeded their targets means that the trading of credits has been limited.

In establishing the TCTP, the TMG provided the demand and supply prospects for the first compliance period, which estimated the demand for credits at about 200,000–300,000 tonnes, while excess reduction credits, including those not yet issued, were estimated at 9.5 million tonnes. By 2016, the owners of 1.3 million tonnes of issued credits were willing to sell them on the market, and those owning 520,000 tonnes were willing to sell, according to a TMG survey (Argus Media 2015). Such excess supply affects the carbon price, which was initially JPY 9600/tonne but fell to JPY 3500 in 2016 (Nishida et al. 2016). Therefore, excess supply will probably continue during the second period and carbon price increases seem unlikely. Overall, the TCTP plays an important role in enabling the mandatory emission reduction programme by providing participants flexibility when taking measures to fulfil their obligation and by demonstrating putting a price on carbon emissions and preparing a full-scale carbon market for the future (Nishida et al. 2016). The trading of credits remains a supplemental measure rather than a driving force. This is obviously different from some other cap-and-trade programmes in which trading plays a much stronger role to lead allowances to their highest-valued use, that is, covering those emissions that are most costly to reduce. However, when Tokyo moves forward to the next stage in carbon emissions mitigation, the ETS scheme will be crucial to fulfil the target.

In the context of establishing a potentially controversial and mandatory programme, political support proved to be critical for the staff of the Bureau. Implementation and governance issues relating to the scheme are important factors in the themes of this book. Governor Ishihara gave his commitment to the team that he was determined to take a strong position in responding to climate change. The experience of the Bureau in developing and implementing the cap-and-trade programme is revelatory in terms of the principles of the performance management doctrine. The clarity of goals to reduce emissions and the willingness to adopt a regulatory approach were particularly important in reducing the emergence of conflicting objectives. Senior managers maintained a high degree of autonomy in decision-making throughout the development of the scheme (Fig. 7.6).

The programme benefitted from a commitment to use a high-quality performance measurement system that provides data for consultation, policy learning, and reporting. The application of an integrated management style, supported by performance data, has been critical for evidence-based decision-making in the Bureau. The practices and procedures of the Bureau helped establish trust between the TMG and the building owners and operators, which became a critical factor to the programme's overall effectiveness in reducing emissions. There are lessons for other city governments from the experiences of the TMG with the various aspects of the performance management framework established in Chap. 4. The next section outlines this experience through data gathered from interviews with TMG staff in November 2016.

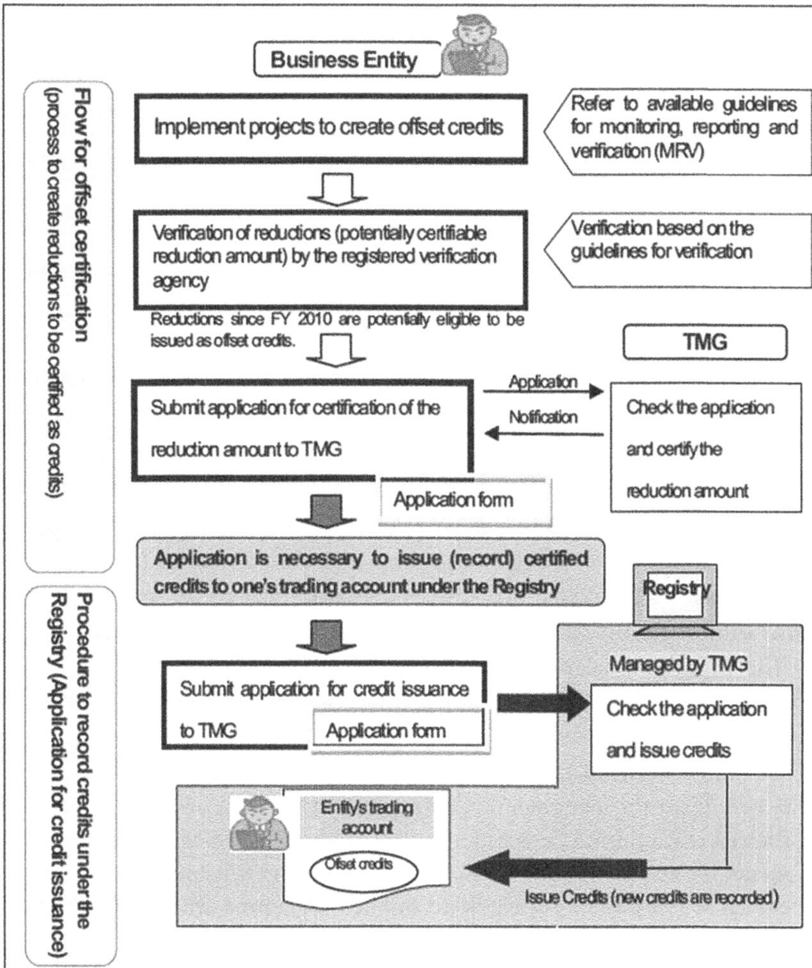

Fig. 7.6 Operation of the TMG cap-and-trade scheme
Source: TMG (2012)

Performance Management and the TMG Cap-and-Trade Scheme

Data collected by the Bureau throughout the mandatory reporting period since 2005 established base line and base emissions information to be allocated to each of the 1300 buildings. The power of this collected data and performance information permitted the specified facilities to make plans in the short, medium, and long term to achieve emission reduction goals by 2020 (Roppongi et al. 2016). It also provided the Bureau with a database of good-quality data based on actual emissions from these buildings, including direct measurement based on fuel/electricity/gas purchase data, that could be used for setting targets and establishing a measurement scheme for the implementation of a climate mitigation policy focused on buildings (personal communication, June 2017). The revision of the Environment Ordinance in 2005 established a mechanism to provide guidance for emissions reduction plans and an evaluation system to reward high performers. The voluntary nature of the initial arrangements, however, resulted in a lack of serious progress towards the 2020 goal. Senior managers in the TMG Environment Bureau recognised the shortcomings of the voluntary programme and took a proposal to the Governor in 2007 to establish a compulsory scheme (Roppongi et al. 2016). With his support the Bureau began a process that would lead to the introduction of a set of municipal ordinances that would focus on reducing emissions in the city by 2020.

The mandatory cap-and-trade scheme was proposed in the *Climate Change Strategy 2007*. The senior planner on the project explained that:

> We thought a mandatory scheme was going to be critical to establish a fair environment for investments. Our experience with the voluntary scheme showed it would not be enough to support large and longer-term investments for emission reductions. With a mandatory scheme there is some equity, as every facility is regulated by the programme. They all share the cost for reductions and therefore there is a level of fairness to the conditions for investment. (Personal communication, November 2016)

The TCTP is an indirect system as its focus is on energy consumption not production. Electricity is generated outside the city, primarily from coal- and oil-fired power stations, which make up over 60% of supply. Constraints on the expansion of the electricity supply sector since the decision by the national government to close some nuclear plants following the Fukushima nuclear disaster have raised concerns in Japan and changed the energy supply structure, forcing the identification of new ways to use existing supplies more efficiently and effectively (Zhang et al. 2012). Renewable sources are becoming increasingly important considerations as over 90% of fossil fuel supplies need to be imported (FEPC 2015; Kuramochi 2015). One interviewee claimed one of the Bureau's key objectives had an economic focus:

> We were aware we needed to help create a good environment for investment in energy conservation and renewable energy. Without being clear on this objective it would have been more difficult to get agreement from stakeholders, and to prompt steady and significant emissions reduction in this sector. (Personal communication, November 2016)

Senior managers in the Bureau were in agreement that the TCTP needed to be as simple as possible so as to reduce confusion and prevent resistance. One manager made the following observation:

> Other cap-and-trade systems have been criticized for their complication with their exception clauses and trade off rules. The broader coverage you have in a scheme means you need to create complicated rules and exemptions. We needed simplicity in our scheme and that really helped our stakeholders to understand the programme and influenced their perception on its fairness. (Personal communication, November 2016)

Throughout the 12-year reign of Governor Ishihara, the staff in the Bureau were assured of his support in the pursuit of climate goals. Many managers could also recollect the period in the 1970s–1980s when the TMG was a leader in Japan for implementing environmental sustainability initiatives. The experience of that period and the political leadership provided by the Governor would prove to be instrumental for climate policy in the city.

Political Leadership and Goal Clarity

The role of Governor Ishihara was crucial throughout the development and implementation of the TCTP, within the TMG, with local stakeholders, and in discussions with the national government. The Governor was a confrontational and often controversial political leader throughout his period in office (Aoki 2014). He remained committed to climate change objectives and attempted to gain the support of leaders from other prefectures and the national government (JFS 2008; Soble 2008). Through Ishihara's leadership the Bureau built on its legacy of being progressive on environmental issues in Japan (Ohno 2010). The widely held view that the national government lacked a coherent and comprehensive approach to climate policy (Suwa 2009; Roppongi et al. 2016) created the opportunity for both the Governor and the Bureau to demonstrate their capacity to take a leading role and attract global attention.

Goal clarity supported by the political leadership played a crucial role in policy development within the Bureau. The clear target of a 25% reduction in emissions specifically focused on buildings in the city made it clear for the senior Bureau managers how they should '*focus their analysis and proposals in policy design*' (personal communication, November 2016). Similarly, a willingness by the political leadership to adopt a long-term perspective in attempts to reduce emissions allowed managers to gather data and design a set of proposals with some security, and proved to be invaluable to the implementation of the climate change strategy. Staff in the Bureau worked closely with the Governor in determining the reduction targets. One senior manager claims:

> Our targets were set by what evidence showed was achievable with the regulatory powers of the city. The Governor's main requirement was that we had to substantially reduce emissions—we identified what was ambitious and achievable and lucky for us the Governor agreed. (Personal communication, November 2016)

The Governor maintained a strong position in his negotiations with the national government and with his own political party, so the opportunity to reveal national weaknesses was important. His reputation for effective climate policy implementation was established with the *Diesel Replacement Scheme*, and the potential for a successful cap-and-trade scheme proved to be attractive for strengthening his credibility. Senior planners in the Bureau argue that:

> the implementation of the Diesel scheme made it clear to us that the mandatory approach was going to be more effective in changing community behaviour than making recommendations or offering incentives for people to change. (Personal communication, November 2016)

This was an important lesson for their work on other climate initiatives. Success with these schemes helped establish what one planner called a *'culture for regulations within the Bureau, because at the time this was the preference at the political level'* (personal communication, November 2016). Senior staff in the Bureau argue the development of the TCTP was reliant on the factors that were present in the TMG at the time, largely due to the leadership style of the Governor. These factors were also important in the application of performance as a means of promoting management and policy change that helped to push the scheme along. The process of developing the climate strategy had taught them the power of good-quality data in developing convincing arguments for both politicians and other stakeholders. According to one senior officer:

> there is an inbuilt resistance to change in some of the TMG Bureaus and we realised that if we were going to push the mandatory approach to the climate agenda, that included a performance framework, then we needed to approach these managers with strong arguments backed up by evidence and clear figures to reduce the possibility of misinterpretation. (Personal communication, November 2016)

Measurement and the Use of Performance Information

The efforts of the Bureau to establish performance data on the activities of commercial and industrial facilities during Phase 1 of the *Environment Security Ordinance* in 2000 proved to be a fundamental aspect in working with stakeholders and designing the emissions trading scheme.

> The TMG had a decade-long dataset to analyse industrial activities and existing reduction plans, which helped to create a detailed policy design that matched local conditions. (Roppongi et al. 2016, 11)

Factual data, provided by building managers and engineers, based on actuals rather than proxy information, proved critical to constructive discussions with stakeholders. The quality of the data sets on energy consumption date provided to the TMG was ensured through a verification process, required by ordinance, undertaken by independent registered agencies (personal communication, June 2017). In meetings with the Bureau team, stakeholder groups who originally objected to the scheme were faced with well-prepared counterarguments supported by high-quality data that highlighted the potential benefits of improved efficiencies in energy use and corresponding emission reductions (Rudolph and Kawakatsu 2012). Since 2005, the TMG has been building strong relationships with building facility managers and engineers through regular discussions on energy efficiency and cost reductions. Bureau officers were experienced in undertaking a consultative approach and recognised the need to improve understanding of the programme with quality data as well as open discussions about enforcement. One manager claimed:

> We work very closely with our stakeholders. With the cap and trade scheme we were particularly concerned to get the technical issues examined by engineers and legal issues identified by our lawyers. The addition of the performance based scheme made us highly sensitive to identifying potential problems. (Personal communications, November 2016)

Data relating to energy efficiency and emissions from the buildings has been based on 128 basic measures developed by the TMG. Building owners have been required, by ordinance, to submit the checklist that shows which measures were taken and to what extent in each building. There is also feedback provided by the TMG to the building operators and engineers outlining good performance and poor performance in terms of reducing emissions based on the experience of others. Such advice and support helps to maintain strong relationships between the TMG and the stakeholders. The participation of the public in stakeholder meetings has also provided the TMG with opportunities to use data in a way that supported public interest questions and to help raise the level of trust in the Bureau decisions on climate issues by exposing potential self-interest from building owners and property developers. The objectives of the

TCTP have been reinforced through working closely with NGOs from the environment sector. According to one manager, this alignment is part of the standard approach taken by the Bureau:

> We have an established practice of consulting and testing our ideas with the NGO's, like WWF and the Japan Climate Network, it's something we have been doing since the 70s and 80s. (Personal communication, November 2016)

Since 2010, the Environment Bureau has been using performance data to support an ongoing policy learning programme. This enables staff to provide a high level of service and advice to the building owners under the TCTP. When combined with regular site visits to the facilities subject to the scheme, they maintain close contact with these stakeholders and provide suggestions for 'improvements to measures and supervise energy reduction initiatives' (Niederhafner 2013, 18). As part of the implementation process, information such as the annual emissions, also verified by independent registered agencies, from facilities and the annual compliance status of facilities as a whole is available to the public online through an electronic database operated by the Bureau (TMG 2012, 30). The view within the TMG is that facility managers and operators are increasingly aware of the power of the application of performance data, diagnosis reports, and benchmarking data. TMG staff claim that through the scheme, both sides have been exchanging practical views on reduction measures for buildings of various types and conditions (Nishida and Hua 2011). According to the Bureau, the availability of high-quality data to both internal and external stakeholders has been 'very effective in promoting the programme' (Nishida and Hua 2011; 531). According to a TMG survey of building owners, managers, operators, and tenants, since the implementation of the TCTP there has been an increased awareness of energy efficiency and the need to reduce energy use. Survey results show that 72% answered that the level of interest from senior management has increased, 75% answered that the motivation of those who engaged in energy savings has risen, and 73% answered that it has become easier to gain cooperation from employees in their facilities (Nishida et al. 2016).

The mandatory aspect of the scheme is supported by the performance monitoring system, which enables the Bureau to notify facility owners to improve compliance performance when they fail to report disclosure, planning, and maintenance obligations (TMG 2012). The fact that the city has established 50 air quality measuring stations across the city since the 1960s means they have substantial trend data that can *'pretty accurately measure changes in different areas'* (personal communication, November 2016). The Bureau participated as a pilot city in the development of the C40 community measurement protocol; this helps to facilitate benchmarking progress against other cities and keeps the performance measurement system up to date.

In circumstances where the verification process reveals that individual building targets have not been achieved, the Bureau has the legislative power to order the facility to reduce its emissions. The penalties for non-compliance include a fine of up to JPY 500,000, including public exposure, a 'name and shame' approach of the breach, and an additional fee to purchase reductions credits. While performance information has been used with some success in working with local stakeholders and for internal management purposes, reporting to the wider community seems to have a lower focus for the Bureau. There has only been one formal report on the progress of the Climate Strategy since 2010 (TMG 2010). TMG officers also claim there are regular media releases that outline emissions reductions and the progress of the scheme (TMG 2016a). Some select data is also available through the city government website or produced in the TMG annual report.

Administrative Leadership and the Incorporation of Performance

The relationship between strong political leadership and management groups within the Bureau committed to the mitigation objectives of the climate strategy proved to be important in the implementation phase of the scheme. Starting in 2000, the Bureau management focused on the development of a high degree of technical competency to establish a capacity for effective implementation (Roppongi et al. 2016). Managers in the Bureau recognised a need to develop the capacity to manage complex and controversial climate programmes and began building skills and knowledge within the Bureau by appointing personnel of qualified engineers and energy emissions technical experts. One manger claimed:

> We realised our approach had the potential to raise a lot of resistance in the business community, so we wanted to make sure we had the skills within the Bureau to deal with strong arguments against what we were trying to do. Our experience in 2005, when we mooted the introduction of an ETS, taught us a great deal about what needed to be done and the skills we needed. (Personal communication, November 2016)

With this capacity, the TMG was in a position to evaluate emission reduction plans and give specific advice to improve measures and to provide options for alternative approaches. Planners within the Bureau recognised the unique managerial environment that encouraged the consideration of new ideas and approaches to policy development. One planner claimed:

> I came to the Bureau from another part of the TMG because it had a reputation for wanting to try new ways of doing things. Other Bureaus in the TMG have an entrenched bureaucratic approach that likes process and established ways of doing things, the Bureau wasn't like that at all. (Personal communication, November 2016)

The management style in the Bureau supported a high level of interaction between its various specialised units. Managers knew they had the political backing of the Governor for their work and this reinforced their commitment. The views of the interviewees are reflected with this statement from one manager:

> We knew the Governor was serious about taking a strong position to reduce emissions. This was highly motivating and strengthened our commitment. (Personal communication, November 2016)

Despite this high level of vertical integration within the Bureau, horizontal integration of policy implementation across the TMG has not been as strong. While interviewees were not necessarily critical of other Bureaus within the TMG, they conceded that cooperation could be improved. As one senior planner stated:

> There will always be a clash between the innovative and bureaucratic approaches, and in many ways we need to develop strategies to deal with it, to get our message across. But there will always be a resistance to using performance in the way we do in the Bureau. Many in the TMG prefer not to know whether they need to improve the way they do things. (Personal communication, November 2016)

Technical skills within the Bureau are juxtaposed by communication and negotiation skills crucial in establishing involvement and support from stakeholders. While arguments may not necessarily work across the different Bureaus in the TMG, they have been effective with external stakeholders in the development and ongoing performance of the cap-and-trade scheme. The strategy of encouraging the involvement of stakeholders in the design of the scheme helped to obtain legitimacy and support for the climate policies. The integrated approach within the Bureau influenced the consultation process with stakeholders. The effectiveness of the consultation process in canvassing objections and creating and modifying operational rules left little rationale for capped facilities to object to or oppose the mandatory system. The Tokyo Chamber of Commerce and Industry and the national Keidanren initially objected to the fundamentals of the scheme. However, the consensus based approach, with strong political backing, supported members of the Environment Bureau in their negotiations with the industry groups, and eventually established agreement from the business sector (Nishida and Hua 2011).

> By involving both opposing and supporting members in the formal stakeholder meetings, the Bureau was able to present critical issues and respond to concerns of the business stakeholders, while supporting arguments were also presented through the voices of civil society organizations. (Roppongi et al. 2016, 11)

Researchers argue that the role the Bureau plays as a councillor, activator, and motivator to support the building owners and operators in their compliance efforts has been instrumental in the negotiations with industries such as the property developers (Niedenhafner 2013). It has been the attention to detail using performance data and the close cooperation and regular and long-term contact between the relevant personnel in administration and businesses, as well as a level of trust between the participants, that allows the Bureau to steer the operators more effectively towards compliance.

Conclusion

Consistent political support from the Governor for over a decade has been a determining factor in building the capacity of the Bureau of Environment to undertake ambitious climate initiatives. The willingness to adopt a regulatory approach has been critical for the clarity of the objectives of the *Green Building Programme*, the *Diesel Reduction Plan*, and the *Cap and Trade Scheme*. The relevant ordinances provided the political legitimacy to support the compulsory application of performance and assessment. The nurturing of in-house skills provided a unique organisational culture within the Bureau that focused on innovative approaches supported by the use of evidence from performance measurement to promote policy and management change. Performance information has been used in policy development and implementation when Bureau staff worked with external stakeholders including adjoining municipalities, industry groups, and building owners. Performance reporting reveals the success of the programmes by providing evidence of reductions in emissions and improvements in air quality across the city.

The lessons from the last decade will provide a substantial body of knowledge as the TMG moves to focus on other sources of emissions and considers how to approach the challenges of climate adaptation. New challenges will emerge as the Bureau shifts its focus to smaller private enterprises that make up the bulk of buildings in the city. A low level of horizontal incorporation of performance measurement across the different Bureaus could be a long-term barrier to a more comprehensive set of climate initiatives, particularly adaptation which requires contributions from land use planning, infrastructure development, and water management. Of greater concern is the change in the political agenda as new Governors elected since 2012 have in each case not been willing to pursue a regulatory approach. The 2016 *Tokyo Environment Master Plan* re-establishes the voluntary approach towards sustainability policy. For example, the TMG will attempt to achieve greater energy efficiency in the city by encouraging the use of electric and hybrid vehicles and promoting bike-sharing (TMG 2016d, 3). The successful bid to host the 2020 Olympics in Tokyo has required a sustainability focus for projects associated with events. Hosting the Games has introduced a new political dynamic to the city that concentrates on the linkages between sustainability and economic opportunity. The objective is to 'realize a world-leading, environment-conscious city positioned to develop and grow in the future' (TMG 2016c). Whether this new competitive dynamic impacts on the ongoing implementation of the climate agenda remains to be seen. What is reasonably clear is that a voluntary approach will provide new challenges for the application of performance as a means of measuring whether encouragement and incentives will deliver the outcomes the TMG claims it is seeking.

References

Adachi, S. A., Kimura, F., Kusaka, H., Inoue, T., & Ueda, H. (2012). Comparison of the Impact of Global Climate Changes and Urbanization on Summertime Future Climate in the Tokyo Metropolitan Area. *Journal of Applied Meteorology and Climatology, 51*(8), 1441–1454.

Aoki, M. (2014). Controversial to the End, Shintaro Ishihara bows out of politics. *The Japan Times*, December 16.

Argus Media Ltd. (2015). Trading Price Valuation Result of Trading Price in Mandatory Total Emission Reductions and Emission Trading Program (TCTP). Retrieved from http://www.kankyo.metro.tokyo.jp/climate/large_scale/attachement/151118shiryou5.pdf

Berinato, S. (2008). Tokyo's 'Just Do It' Leadership on Climate Change. *Harvard Business Review*, June 3.

CLAIR (Council for Local Authorities for International Relations). (2016). Fact Sheet: An outline of Local Government in Japan. https://www.gdrc.org/uem/observatory/logov-japan.pdf

EDF (Environment Defence Fund). (2015). *Tokyo: An Emissions Trading Case Study.* https://www.edf.org/sites/default/files/japan-case-study-may2015.pdf

FEPC (Federation of Electricity Power Companies of Japan). (2015). Japan's Energy Supply Situation and Basic Policy. http://www.fepc.or.jp/english/energy_electricity/supply_situation/

Fujibe, F. (2011). Urban Warming in Japanese Cities and Its Relation to Climate Change Monitoring. *International Journal of Climatology, 31*(2), 162–173.

Goto, K., Yabe, K., & Araki, N. (2004). *Estimation of Maximum Demand Sensitivity to the Temperature for Long-Term Power Demand Forecast*. Proc Technical Meeting of Power and Energy Division.

Hijioka, Y., Takano, S., Oka, K., Yoshikawa, M., Ichihashi, A., Baba, K., & Ishiwatari, S. (2016). Potential of Existing Policies of the Tokyo Metropolitan Government for Implementing Adaptation to Climate Change. *Regional Environmental Change, 16*(4), 967–978.

Hooper, R. (2013). Effects Will Become More Obvious as Japan's Climate Changes. *The Japan Times*, July 13.

Horiuchi, Y., Saito, J., & Yamada, K. (2015). Removing Boundaries, Losing Connections: Electoral Consequences of Local Government Reform in Japan. *Journal of East Asian Studies, 15*(01), 99–125.

Hosomi, M. (2010). What Might the Environmental Policy of Tokyo Metropolitan Government Achieve? *Clean Technology Environment Policy, 12*, 451–453.

IEA (International Energy Agency). (2008). *Energy Policies of IEA Countries: Japan 2008 Review*. Paris: OECD/IEA.

JFS (Japan for Sustainability). (2008). *Tokyo Takes the Lead in Climate Change Mitigation Measures in Japan*. http://www.japanfs.org/en/news/archives/news_id027949.html

JFS (Japan for Sustainability). (2011). Tokyo Metropolitan Government's Efforts to Control Diesel Vehicle Emissions. https://www.japanfs.org/en/news/archives/news_id030817.html

Jones, R. S., & Yoo, B. (2009), *Improving the Policy Framework in Japan to Address Climate Change*. OECD Economics Department Working Papers, No. 740, OECD Publishing. DOI 10.1787/218874608768

Kanai, T. (2015). Local Governments and Policy Analysis in Japan after the Second World War. In Y. Adachi, S. Hosono, & J. Lio (Eds.), *Policy Analysis in Japan*. Bristol: International Library of Policy Analysis.

Kikuchi, M. (2007). Assessing Government Efforts to (Re)build Trust in Government: Challenges and Lessons Learned from Japanese Experiences. *International Public Management Review, 8*(2), 183–203.

Kim, S. (2010). Public Trust in Government in Japan and South Korea: Does the Rise of Critical Citizens Matter? *Public Administration Review, 70*(5), 801–810.

Kimura, K., & Zissler, R. (2016). Renewable Energy Institute: Comparing Prices and Costs of Solar PV in Japan and German – The Reasons Why Solar PV is more Expensive in Japan. Tokyo: REI, p. 25.

Kudo, H. (2015). Bridging the Gap Between Reform Practices and Literatures: New Public Service Delivery of Japanese Local Governments. *International Public Management Review, 16*(1), 147–164.

Kuramochi, T. (2014). *GHG Mitigation in Japan: An Overview of the Current Policy Landscape*. Working Paper. Washington, DC: World Resources Institute. wri.org/publication/ghg-mitigation-policy-japan

Kuramochi, T. (2015). Review of Energy and Climate Policy Developments in Japan Before and After Fukushima. *Renewable and Sustainable Energy Reviews, 43*, 1320–1332.

Kusaka, H., Suzuki-Parker, A., Aoyagi, T., Adachi, S. A., & Yamagata, Y. (2016). Assessment of RCM and Urban Scenarios Uncertainties in the Climate Projections for August in the 2050s in Tokyo. *Climatic Change, 137*(3-4), 427–438.

Lam, P. (2011). Japan's Environmental Politics and Change Local, National and Global. In A. Gounder (Ed.), *Routledge Handbook of Japanese Politics*. London: Routledge.

MLIT (Ministry of Land, Infrastructure, Transport and Tourism). (2014). *Outline of Low Carbon City Development*. City Bureau: City Planning Division.

MOE. (2016). *The Basic Environment Plan*. https://www.env.go.jp/en/policy/plan/basic/index.html

MOE (Ministry of the Environment). (2012). *Consolidtated Report on Observations, Projections and Impact Assessments of Climate Change*. Climate Change and its Impacts in Japan. FY2012. Global Environment Bureau.

Niederhafner, S. (2013). The Governance Modes of the Tokyo Metropolitan Government Emissions Trading System. http://hdl.handle.net/10086/26005

Nishida, Y. (2013). Energy Efficiency in Buildings: Policy and Programs in Tokyo. Presentation. Tokyo Metropolitan Government – Environment.

Nishida, Y., & Hua, Y. (2011). Motivating Stakeholders to Deliver Change: Tokyo's Cap-and-Trade Program. *Building Research & Information, 39*(5), 518–533.

Nishida, Y., Hua, Y., & Okamoto, N. (2016). Alternative Building Emission-Reduction Measure: Outcomes from the Tokyo Cap-and-Trade Program. *Building Research & Information, 44*(5-6), 644–659.

Ogawa-Onishi, Y., & Berry, P. M. (2013). Ecological Impacts of Climate Change in Japan: The Importance of Integrating Local and International Publications. *Biological Conservation, 157*, 361–371.

Ohno, T. (2010). *Low Carbon Metropolis: Tokyo's Climate Change Measures*. Presentation to UCLG Congress 2010, Hamamatsu.

Okamoto, M. (2012). *Municipal Amalgamations in Japan: Who is happy?* Paper prepared for presentation at the 22ed IPSA World Congress, Madrid, Spain, July 8–12 2012, Local Government Amalgamation: Effects on Democracy and Governance, Session: RC05 Comparative Studies on Local Government and Politics.

Roppongi, H., Suwa, A., & Puppim De Oliveira, J. A. (2016). Innovating in Sub-national Climate Policy: The Mandatory Emissions Reduction Scheme in Tokyo. *Climate Policy, 17*, 516–532.

Rudolph, S., & Kawakatsu, T. (2012). Tokyo's Greenhouse Gas Emissions Trading Scheme: A Model for Sustainable Megacity Carbon Markets. *Market Based Instruments: National Experiences in Environmental Sustainability*, 77–93.

Schreurs, M. A. (2010). Multi-level Governance and Global Climate Change in East Asia. *Asian Economic Policy Review, 5*(1), 88–105.

Shimizu, K. (2012). Electoral Consequences of Municipal Mergers. *Journal of East Asian Studies, Sep-Dec, 12*(3), 381–408.

Soble, J. (2008). Tokyo Mayor Leads on Climate, *Financial Times*, June 3.

Sugiyama, N., & Takeuchi, T. (2008). Local Policies for Climate Change in Japan. *The Journal of Environment & Development, 17*(4), 424–441.

Suwa, A. (2009). Soft Energy Paths in Japan: A Back Casting Approach to Energy Planning. *Climate Policy, 9*, 185–206.

Suzuki, Y., Kakei, F., Yamamoto, C., Sawaki, M., Matsunaka, R., & Moritsu, Y. (2015). Studies on the Enforcement of the Low Carbon City Promotion Act and a Case Example of Low Carbon City Development Plan in Omihachiman City. *WIT Transactions on Ecology and the Environment, 193*, 421–432.

Takai, K. (2014). *Sustainable Urbanisation in Japan Aiming Towards a Low Carbon City*. Presentation to the United Nations Centre for Regional Development.

Takahashi, E. (2011). The Diesel Vehicle Control in Toyko. Tokyo Municipal Government. https://www.kankyo.metro.tokyo.jp/en/automobile/attachement/00_tmgs_challenge.pdf

Tanaka, M., Shirai, N., Ogawa, M., Kimura, H. (2011). Regional Impacts of Global Warming, Summarized Results of Fact-Finding Survey (Literature Search) on Efforts Toward Adaptation Policies. http://www.hosei.ac.jp/NEWS/newsrelease/110224.html

TMG. (2008). *Tokyo Metropolitan Environmental Master Plan*. https://www.kankyo.metro.tokyo.jp/en/attachement/Master-Plan(Outline).pdf

TMG. (2010, March). *Tokyo Climate Change Strategy: Progress Report and Future Vision*. Tokyo: Tokyo Metropolitan Government.

TMG. (2011). *The Diesel Vehicle Control in Tokyo*. Bureau of Environment. E. Takahashi presentation. Automotive Pollution Control Division.

TMG. (2012). *Tokyo Cap and Trade Program for Large Facilities*. Detailed documents. Tokyo Metropolitan Environmental Security Ordinance. Bureau of Environment.

TMG. (2016a). Tokyo Cap and Trade Program achieves 25% reduction after 5th year. Tokyo Metropolitan Government, Bureau of Environment. http://www.kankyo.metro.tokyo.jp/en/climate/cap_and_trade.html

TMG. (2016b). Resource Directory: Bureaus and Departments. http://www.metro.tokyo.jp/ENGLISH/DIRECTORY/bureau.htm

TMG. (2016c). *Creating a Sustainable City: Tokyo's Environment Policy*. Tokyo: Tokyo Metropolitan Government.

TMG. (2016d). *Tokyo Environmental Master Plan*. Bureau of Environment. Sustainable Tokyo, Tokyo Municipal Government.

TMG (Tokyo Municipal Government). (2007). *Tokyo Climate Change Strategy: A Basic Policy for the 10 Year Project for Carbon Minus Tokyo*. Tokyo: TMG.

TMG (Tokyo Municipal Government). (2017). The structure of the Tokyo Metropolitan Government. http://www.metro.tokyo.jp/ENGLISH/ABOUT/STRUCTURE/structure04.htm

Tsujiyama, T. (2002). Jichi Kihon Jorei no Kousou (Perception of Local Government Charter). In K. Matsushita, M. Nishio, & M. Shindo (Eds.), *Jichitai no Kousou (4) Kikou (Perception of Local Government, vol. 4, Institution)*. Tokyo: Iwanami Publisher.

UNFCC. (2015, December). *Japan's Second Biennial Report under the United Nations Framework Convention on Climate Change*. Tokyo: The Government of Japan.

Wang-Helmreich, H. (2015). The Influence of Domestic Society, Policy and Politics on International Climate Policy in Japan: The Development of a Country's Positioning in Climate Negotiations under the UNFCCC.

WWF (World Wildlife Fund). (2008). *Nippon Changes: Climate Impacts Threatening Japan Today and Tomorrow*. Tokyo, Japan: WWF.

Yorifuji, T., Kawachi, I., Kaneda, M., Takao, S., Kashima, S., & Doi, H. (2011). Diesel Vehicle Emission and Death Rates in Tokyo, Japan: A Natural Experiment. *Science of the Total Environment, 409*(19), 3620–3627.

Zhang, Q., Ishihara, K. N., Mclellan, B. C., & Tezuka, T. (2012). Scenario Analysis on Future Electricity Supply and Demand in Japan. *Energy, 38*(1), 376–385.

8

Synthesis and Conclusions

We feel that even when all possible scientific questions have been answered, the problems of life remain completely untouched
—Wittgenstein, L. 1921. Tractatus Logico-Philosophicus, 6.52

Introduction

A key point of this book has been that performance management can help city governments achieve their climate goals. Policy effectiveness and success are uncertain and dependent on numerous factors. This realisation should prompt more realistic approaches from both individual city governments and network alliances about the difficulties involved and lead to more modest expectations. The purpose of academic research in this field is to highlight the challenges and provide insights for more realistic expectations. The performance management logic appears to be simple and intuitive. It looks like common sense. Performance management is succinctly described as 'classic good government reform' (Moynihan 2008, 192). The case studies reveal that linking performance management to climate policy can be challenging for city government managers. Political promises create expectations, both within the government and

S. Jones, *Cities Responding to Climate Change*,
DOI 10.1007/978-3-319-64810-1_8

the city community, that solutions will be found, that information and measurement will provide the answers we need. This can be a perilous mix, as international research shows that targets and performance measures alone do not lead to better performance—'it often leads to dysfunctional behaviours such as sub-optimization, target fixation, cheating and lying' (Marr 2009, 271). Managers recognise the need to reduce opportunities for these dysfunctional behaviours through extensive long-term commitment of resources and high levels of autonomy if performance is to act as the director of reform. Analyses such as this book can contribute to challenging expectations by outlining comprehensively and coherently the barriers and opportunities in implementation.

Despite the efforts of leading cities and the support of network alliances to provide guidance and support, individual city governments have the responsibility to transform their own operations into a performance-driven culture. Political commitment and sustained leadership are the key factors that are consistently linked to high levels of performance management (Sanger 2008; Behn 2008; Moynihan 2008; Aylett 2015). The evidence from these case studies reveals complex circumstances for city governments. Factors such as resource capacity, organisational culture, decision-making processes, and organisation structure all play a major role within the city government to facilitate or inhibit the application of performance management.

The application of various elements of the performance management doctrine by the case study city governments has been shown to support considerable benefits. For example, each city set ambitious GHG emissions reduction targets and established measurement systems that supported reporting results to their communities. Each of the three city governments reveals important issues for the application of performance measurement to support management and policy change. Tokyo has been instructive in its successful application of performance in the development and implementation of a mandatory approach to reduce emissions. Stockholm has emphasised the challenges of political tensions within the city to the application of performance as a method of promoting the mainstreaming of climate policy. Copenhagen provides insights into the importance of a structured evaluation process throughout the implementation of climate policy. Here, the evaluation and review process helps to create opportunities for organisational learning, with new ideas for management change as well as for change to assumptions underlying policy objectives.

Along with the individual experiences of the case study councils, there are insights that are reasonably common across each of the cities. Council officers interviewed for this research have been strong advocates for the benefits of measuring results that contribute to the development of evidence to support policy and management change. They are also convinced about the value of performance as a tool for identifying areas where attention can be directed to improve the implementation of climate policy. The cases reveal the critical contribution of political leadership to the work undertaken by officers engaged in responding to climate change. The support of the Mayor/Governor is particularly important in the application of performance as a means of ensuring vertical and horizontal cooperation between administrative units in the implementation of climate policy. Each case also reveals limitations, particularly from institutional arrangements, that can constrain policy implementation; however, there is some optimism by senior managers that the improved application of performance reporting will reduce this impact.

Using the analytical framework, this book has examined the approaches and practices of the case study city governments. Confirmation was obtained through observable evidence in accordance with the dimensions of performance management. The research findings are based on consultation of original policy documents, secondary research information, and in-depth personal interviews with senior managers in each city. The research has validated that the dimensions of the framework provide the depth of analysis needed to confirm whether principles of the performance doctrine have been adopted in developing and implementing climate policy. As a result, the practices of each city considered in the context of the heuristic are detailed in Table 8.1, which illustrates the stage achieved, with insights into the requirements for change that promise greater benefits from a performance-based approach.

These findings are important for the three agendas of city governments (international, engagement, and organisational governance), outlined in Chap. 3, set by city governments nationally in their multilevel systems and internationally as established networks and coalitions to support

Table 8.1 Stages of performance management in city governments for climate policy

Stage	Goal clarity	Measurement	Incorporation	Use
1. Minor involvement	Aspirational, voluntary approach, vague objectives	No measurement commitment, no stakeholder consultation	No specific responsibility for performance, no management autonomy	Minor public reporting vague in specifics
2. Limited commitment	Mandated and voluntary targets in selected responsibilities	Measures introduced in chosen policy areas with limited stakeholder consultation	Incorporate performance in specific functional areas, limited management autonomy	Reporting, internal and external, limited to areas that show progress; single-loop learning
3. Comprehensive participation	Mandated targets in areas responsibility of city relevant to climate action	Specific measures in all areas established with extensive stakeholder consultation	Incorporation of performance across all targeted areas with relevant organisation units, management autonomy	Public reporting in all aspects of policy action designed for relevant stakeholders; double- and triple-loop learning

their actions. In the context of the claims being made by city governments and from the ambitious targets and proposed actions, it is critical that they demonstrate their success through the transparent application of clear and objective measurements and use performance information in accurate and meaningful ways. The results outlined here will help city governments to learn from the experience of the case study governments and gain insights on how to improve their performance management approaches for climate change action.

In the next sections, the chapter will outline the conclusions reached from this research on the contribution of each of the dimensions of the analytical framework. The conclusions will discuss issues present in the city governments in the context of the current literature on both climate policy and performance management. The heuristic provides a map for policy-makers, with markers for establishing progress in improving performance that contribute to policy success. Each stage outlines the status of performance in accordance with the dimensions of performance management examined through the analytical framework. Specific relevance to performance and climate policy will also be featured. The assumption behind this approach is that applying the right principles and doing performance management in accordance with the doctrine will lead to better performance and more policy success and effectiveness by city governments (Marr 2009).

Goal Clarity

The analytical framework argues that goal clarity is the foundation on which performance management systems are built. Clarity takes on added importance for climate policy because of its complex multidimensional nature. Effective climate policy requires contributions from a cross section of city government functions. Without clarity of objectives and expectations, policy actors risk pursuing activities with little impact on city government climate objectives. At its worst, lack of clarity facilitates evasion and gaming by agencies seeking to avoid serious action on climate change. The case study cities presented a mixed picture of goal clarity. In some areas, goals were relatively clear and precise; for example, the

Tokyo city government set the goal of reducing emissions from 1300 of the largest GHG emitters in the city by 25% within a five-year time frame. Rather vague goals were also present, such as Copenhagen's goal of being the world's 'best' city for cyclists. Precise goals allow for precise measures to be developed and clarity for managers and practitioners on the actions that need to be taken and by whom within the organisation. Lack of clarity allows confusion and lack of precision of measures because no one really knows exactly what is to be achieved and how to measure progress.

A key aspect of achieving goal clarity is the practice of strategic planning. The analytical framework argues that a strategic plan, established with stakeholder consultation and agreement, is influential in attaining clarity of goals for city governments. A strategic plan links the objectives in the different functions that are relevant to achieving outcomes. These linkages are critical for the multidimensional nature of climate policy. In the case of Stockholm, the city undertook considerable consultation across its management structure and with external stakeholders in establishing the climate plan for the city. Research on other public sector agencies suggests that the Stockholm approach may be unique. Surveys of government agencies show that 61% of senior managers, 21% of middle managers, and 2% of remaining employees feel engaged in the development of strategy in their organisations (Marr 2009, 274). Stockholm's approach strengthened in 2016 with consultation and sign-off from senior managers required before the new climate plan was presented to the city council for endorsement. This approach raises awareness across city government units and requires commitment from individual managers that climate goals are incorporated into the operational plans under their responsibility.

Evidence from comparative research suggests that city managers have been reluctant to link performance measurement to strategic planning goals (Ammons 2015). Managers from each case study city responsible for climate action expressed strong support for this connection. Managers from the Copenhagen Bike Secretariat argued that this linkage had been a strong tradition to promote cycling through a performance-based approach. In Tokyo, city managers argued that without a performance measurement approach, their work with external stakeholders would not have been an effective contribution to the climate objectives. These connections were an important component in their reporting system and their capacity to demonstrate policy success.

Clarity also helps reduce the possibility of objectives that are at cross purposes or even contradictory. Climate change policy is different from the traditional focus of city government because of its novelty and cross-cutting nature. By deciding climate objectives at the highest levels, councils can prevent contradictory objectives. Policies at cross purposes place senior managers and practitioners in a difficult circumstance when needing to decide which objectives take precedence. In instances such as these, good performance in one can contribute to poor performance in the other. Although the case study cities were quite clear in many of their goals for climate policy, goal ambiguity provided challenges for managers. One representative example was the conflict that emerged in Copenhagen between climate goals and the goal of promoting economic competitiveness to facilitate employment opportunities. There has been a strong economic focus in the city that has seen sustainability principles challenged. Research confirms that the District of Orestad mega-project has fallen short of its sustainability objectives in an attempt to attract development (Loerakker 2013; Book et al. 2010; Majoor 2014). Like other major development projects in large cities, the Orestad development involved a partnership arrangement between the city government and the private sector. Financial difficulties caused by the global financial crisis caused some readjustment of the original goals of the project. Examples such as Orestad highlight the problems city governments can face when there is conflict in partnership arrangements between the government's objectives to promote sustainability and the private sector's need for profit and speed. Governments set up a situation where there is conflict between their role as regulator and their objective to promote development as an investment partner.

Lack of clarity has contributed to challenges within the city administration in Stockholm, where resistance from the transport department has impacted on the achievement of climate goals. Despite the clear goal of improving cycling infrastructure in Stockholm, the transport department has resisted interference with its traffic goals, and cycling infrastructure continues to be a retrofit arrangement after the priorities of road transport have been covered (Koglin 2015). In this case, the complexity of the electoral system and the opportunities of minor issue-based parties also exerted influence through the committee decision-making processes. This influence has served to override the potential benefits of the integrated management system. In this regard, the management approach adopted in Copenhagen, which sees a merging of related functions into the committee structure, has produced a coordinated approach across planning, environment, and transport functions. The question of the effect of decision-making structures on performance management of climate change objectives is one area that warrants more in-depth research in future.

To the extent that clearer goals and roles contribute to organisational performance, city government practitioners strive to understand the mechanisms designed to maximise goal and role clarity. For city governments serious about climate change, the networked nature of policy implementation is unavoidable, limiting the tensions between networks, and public managers' favourable interpretation of their organisation is a critical area of research in public management. Clarity is needed for the individual employee performance schemes. City government staff need to know how their performance will be assessed, and clarity of the objectives of the organisation is fundamental for equity and fairness in the employment contract. Establishing clear goals has implications for the relationship between local politicians and the senior managers and practitioners within the city administration. Administrative structures shape the bureaucratic process, performance, and responsiveness with municipal governments (Krause et al. 2014). Clarity of goals establishes the foundation for the development of targets and measurement protocols that provide the guidance necessary to achieve climate change objectives.

Measurement

Measurement is the dimension of performance management that provides the data for policy-makers to undertake comparison, prediction, and evaluation (Sterck and Scheers 2006). The analytical framework argues that the foundation for measurement is setting achievable targets. A critical aspect of the performance doctrine is that measurement information should contribute to learning, innovation, and creativity within the organisation. Appropriate targets and indicators are vital components to achieving this objective. Urban-scale targets such as whether a city government has the resources, legislative authority or influence, and financial capacity to meet are critical factors for target setting. The case study councils were very careful when setting targets in areas where they were serious about taking climate action. Managers in Stockholm and Copenhagen argued that their process in establishing their climate strategies was to identify areas where they could actually do something to produce a climate outcome. Stockholm managers stressed that they only invited consultation and stakeholder feedback in areas of city responsibility. They did not want to 'open the floodgates' to every aspect of climate change of concern to the community, thereby creating false expectations that the city could influence change in national and international approaches. In this way, their targets would only relate to municipal responsibilities. The participation of stakeholders contributed to developing relevant measures, that is, measures that reflect community concerns about climate action within the context of municipal responsibilities. These measures were vital in establishing reports that provide information to stakeholders about the progress in the implementation of policy.

The analytical framework argues that city governments need to develop good indicators that show linkages between policy action and climate outcomes. This has been shown to be a significant problem in public sector organisations generally. International survey data shows that only 15% of public sector managers see a linkage between their performance indicators and the strategy of their organisation (Marr 2009). This can often be a result of the dominance of performance requirements of other levels of government establishing a compliance-based approach from municipalities to funding arrangements. Research shows however that good performance management occurs when it is viewed as an organisational tool, not as a compliance burden (Moynihan 2008, 198).

Managers from the case study councils presented an optimistic view that, for climate change policy, their performance indicators were of direct relevance to their climate strategies. This may be a result of operating at the local level, being close to their communities, and therefore having a great deal at stake. City managers also work closely with their political leaders and are responsible for both policy development and service delivery. City managers argued that considerable time and resources had been allocated to ensure direct relevance of what was being measured and what was to be achieved. Each city had long-standing measurement resources that enabled the demonstration of results. The capacity of Stockholm to monitor car movements in and out of the city enabled it to demonstrate the impact on traffic of the congestion charge. Tokyo too was able to illustrate the results of its cap-and-trade scheme through simple diagrams that revealed how quickly emissions in the city were being reduced from the buildings targeted as a result of the scheme. Each of the cities established baseline data using the UN Protocol, thus enabling comparison with other cities in the ICLEI and C40 networks. These indicators provide some level of transparency of actions taken and reveal that local politicians are keen to show results and compare them with other cities. The results however are limited because they are largely focused on the production of GHG emissions, not consumption-based emissions in each city. Only Stockholm confirmed any current interest in developing the indicators focused on consumption-based emissions as an area for future examination.

Institutional arrangements mean that the case study councils have access to considerable financial resources through tax revenue arrangements that give them the capacity for collecting and processing measurement data. This is not the situation for many smaller city councils in other contexts, where data collection can be challenging due to resource constraints. The resources available to Stockholm, Copenhagen, and Tokyo provide them with the capacity to build and manage performance information systems. The seriousness with which Copenhagen has approached its policy to reduce emissions by increasing cycling in the city produced comprehensive, in-depth reports since the 1990s, which are being improved continually. The coupling of cycling with city branding and economic competitiveness strategies, as well as with GHG reduction objectives, helps to ensure new measures are developed, reporting is under continual review, and the allocation of considerable resources is dedicated to data collection. Interestingly, reporting on other aspects of Copenhagen's CPH 2025 strategy does not receive the same level of attention or promotion as cycling.

Managers with responsibility for climate policy also need to not lose sight of the environmental indicators particularly when economic matters dominate political attention. One example is the application of the congestion charge in Stockholm, serving as a constraint on the consideration of a broader set of ideas to achieve the same objective. The focus of the measures will determine the focus of the management attention. 'What gets measured, gets managed' is Drucker's well-worn axiom. While the introduction of the congestion charge in Stockholm produced a reduction in car travel, this was largely based on economic factors, such as costs to drivers and revenue to the government, rather than on a recognition that car travel contributes to GHG emissions and should be significantly reduced for ecological reasons. The fact that car travel plateaued after the introduction of the congestion charge has established the view in the city government that further reduction in traffic volumes will only be possible with increases in the cost of the congestion charge. So, rather than behaviour change for environmental or climate reasons, the indicators have been linked to economic cost factors.

The reality for city governments is that they are highly competitive in attracting investment. The public performance reports examined for this research are designed specifically with this in mind. As mentioned in Chap. 2, the 'green agenda' falls nicely into the pro-development approach many cities are pursuing to attract investment. The adoption of entrepreneurial city marketing approaches utilises performance information in a competitive way that is 'fundamentally about comparing, contrasting, and ranking cities for specific purposes' (McCann 2013, 7). Local politicians understandably want to measure progress and dynamism, and climate reporting needs to be considered in this context. City governments are willing to benchmark their result with what they identify as their competitors. This approach was particularly evident in the case of Copenhagen and Stockholm, where investment attraction in highly competitive in the European context. The focus of measurement for these cities will therefore be external and across a broad cross section of areas, but the reports will not be in depth or comprehensive. Managers of climate change programmes need to look for ways to protect their initiatives from criticism and maintain resource commitments. Linking indicators to results from other city government priorities could help entrench climate initiatives. Tokyo provides an interesting example of how circumstances can change the nature of data collection and indicators of progress. Early reports on the progress of the climate change strategy were quite detailed, covering most of the areas of policy action. This reflected the Governor's focus on climate action and his/her lack of concern for competing with other cities. The Governor's concern was to look for ways to embarrass the national government into action on climate change by demonstrating the success of Tokyo's approach. More recent indicators reflect the interests of the new, more conservative regime in Tokyo, with reports recently taking a stronger marketing focus as the city government attempts to benefit from hosting the Olympic Games in 2020. As a result, reports in Tokyo are now more sporadic media announcements promoting the city's green credentials.

Measurement and data collection activities should focus government activities towards the achievement of outcomes. The performance management doctrine provides the principles that give governments the tools for engaging in policy and management change. It is reasonable to conclude that the observable evidence identified through the application of the analytical framework provided the analysis of measurement details and the performance indicators in the climate reports. The analysis shows that the indicators reveal very little about management within the city governments. One of the principles of the performance doctrine is that there needs to be autonomy on the part of managers to set targets and design indicators. The commitment of managers is essential to developing creative solutions to the challenges posed by climate change. Not one of the case study cities has been able to achieve strong commitment to climate goals across its organisational units. Entrenched bureaucratic processes and procedures were a common feature that resisted performance measurement and a willingness to be creative and innovative. Not even the highly regarded integrated management system (IMS) in Stockholm was able to overcome political and management resistance to climate goals if other functional areas believed the goals worked against their own objectives. Surveys of public officials suggest that a positive learning culture is more likely when there is strong internal focus on the application of performance data (Marr 2009). The development of contextualised performance measures that are relevant to individual units helps to develop an environment for good performance management. The common practice of data aggregation reduced the opportunity for different units within the case city governments to be held accountable for their actions towards achieving the climate goals. It also highlighted a need for a new approach within these city governments to develop measures and indicators through the involvement of other parts of the organisation so that new ideas can merge and contribute to achieving outcomes.

While innovation may be difficult to observe in public sector agencies, well-designed indicators should provide the details to help managers learn and improve performance. Opportunities to avoid exposure of poor performance, as the experience of Stockholm shows, can restrict opportunities for learning for improvement. A new approach to developing measures and indicators being adopted in Stockholm shows the desire to improve performance in achieving the climate goals in the city. City governments need to develop internally focused performance measures, with indicators that are clear, realistic, and achievable. With improvements to internally focused performance reports that are relevant to managers, there may be a willingness to report more accurately on performance. By allowing managers to experiment with measurement and innovation, there may be an improvement in the level of engagement in measurement design and target-setting processes (Marr 2009). Such benefits will contribute towards embedding performance management throughout the city government organisation and support mainstreaming climate consideration in relevant policy areas.

Incorporation

According to the conceptual framework, performance needs to be embedded into management processes and procedures so that it can be used in effective ways to measure results and support change. As discussed in Chap. 4, incorporation is the link between measurement and use of performance information (Van Dooren et al. 2010). By incorporating performance into their operations, city governments are demonstrating a serious commitment to using it as a tool to support policy implementation and accountability. Incorporation of climate considerations into the performance requirements of an agency or organisational unit within a city government will be critical to the achievement of the climate objectives established by the city policy-makers. Each of the case study cities established coordination units within the environment function of the city administration. These units have responsibility for implementing climate policy, monitoring progress, and reporting results to city politicians. Due to its cross-cutting nature, climate policy requires different approaches to usual practice in the functional areas, such as planning, infrastructure development, transport, and water management. The mainstreaming of climate policy across these functions is an effective way to achieve coherent and operational approaches to implementation and fulfil the council's climate objectives. By incorporating performance against the climate objectives into the management practices of the relevant organisational units, managers will be required to report against progress for their component of the overall climate strategy.

Commitment from across the city government administration is required to maximise the benefits of performance management. For city governments, both politicians and senior managers must be committed to performance management, to get a system designed and implemented. Middle managers and staff need to be committed to make performance management work as part of the processes and procedures to ensure performance information is used and continues to be of value. Interviewees in each case study argued that it was difficult to obtain the level of commitment required to both climate policy and the performance requirements from other functional areas. A common problem in public sector organisations is the declining level of awareness of performance requirements by the lower levels of management and staff. International surveys of the public sector confirm that the level of awareness, acceptance, and use declines in the lower levels of the organisation (Marr 2009). The analytical framework in this book argues that 'street-level' bureaucrats, as discussed in Chaps. 3 and 4, have considerable influence on policy outcomes, and their commitment to successful implementation is a critical element of climate policy. Senior managers in Copenhagen and Stockholm emphasised their efforts to raise awareness about the city government's climate policy within their organisations. Linking both performance and climate policy in the early stages of the development of strategic and operational plans and throughout the budget process helped to embed performance against the climate objectives across the organisation. Managers in Tokyo argued that the level of commitment to climate objectives in other functional areas was an issue that needed new approaches to help foster change. In each case, efforts were being made by city managers to improve their situations. Change is a long-term process requiring vigilance and commitment from both political leaders and senior managers. Without improvement, the benefits of mainstreaming and incorporation of performance will be problematic.

According to the performance doctrine, embedding performance in management processes helps to ensure accountability by each relevant organisational unit for achieving the city's climate objectives. Reluctance by functional areas to pursue their required contribution will not only jeopardise policy success but could raise questions in the community about the level of commitment to climate objectives by the city government. The case studies have revealed that individual units within the council structure are capable of resisting their full contribution to climate goals for reasons ranging from political to administrative. Within the public sector, the traditional organisational culture has emphasised bureaucratic values of hierarchy, process, and procedure (Parker and Bradley 2000; Rainey 2009). Encouraging the innovation and creativity required for climate policy has been a challenge for the case study city governments. As an example, in Tokyo, the traditional bureaucratic management style of different bureaus constrained the adoption of more flexible and innovative approaches needed to meet some of the climate objectives established by the Environment Bureau. As a result of the resistance, implementation was slowed and there was some doubt if some of the climate objectives would be achieved within the specified time frames. According to the analytical framework, embedding performance into the management processes across the organisation helps to identify the source of resistance and to possibly identify ways to resolve it.

Structural factors also play a role in the way climate and performance can be incorporated into decision-making. The organisation structure in Copenhagen plays a significant role in helping to promote cooperation and coordination between functions relating to climate change. Copenhagen has the Technical and Environment Division responsible for transport planning, urban planning, and environmental issues. There is a strong focus on mainstreaming, as managers of each of these functions need to work together to achieve the objectives of the council under the responsibility of one deputy Mayor. In Stockholm, the structure of the organisation is more typical of municipal governments, which tend to separate these functions. In Stockholm, cooperation among the three key functions of urban, transport, and environmental planning has been attempted administratively through the IMS, intended to eliminate conflicting objectives and to ensure sustainability is a consideration in decision-making across the organisational units. The results of the IMS have not been as effective as the managers from the Technical and Environment Division had hoped. Political resistance through the committees system also managed to override the anticipated benefits of the IMS. The lesson here is that mainstreaming is a complex task, and city governments need to consider the impact of organisation structure, political resistance, and decision-making processes to identify potential challenges to incorporating both climate policy and performance expectations.

A core principle of the performance doctrine is the need to promote management autonomy. As outlined in Chap. 4, the assumption is that stronger commitment to organisational goals can be achieved by strengthening management autonomy to establish performance-based processes and procedures. Public sector research suggests that performance information requirements of centralised units are generally seen as a compliance burden, rather than as an organisational tool for improvement (Moynihan 2008). The positioning of a programme within the bureaucratic structure of government—whether as an independent unit or a subpart of another, or whether as a component of the executive or legislative branch—can meaningfully affect its priorities, the amount of resources and authority it has available, and the types of interests that have access to it (Krause et al. 2014, 113). Each of the city managers in the case studies with responsibility for climate change discussed the challenges in gaining cooperation and support from other parts of the organisation in achieving climate goals. For example, transport departments regarded the addition of bicycle lanes to their infrastructure projects as an inconvenience and an additional cost. Recycling and waste management in accordance with climate and sustainability objectives created challenges for waste management units. Departments often regarded climate initiatives as competition for limited council funds. The performance literature reinforces the need for decentralised approaches to performance management. With decentralised approaches, individual managers have greater opportunities to identify ways in which they can contribute to climate objectives. Managers play a pivotal role in using performance information to identify ways to improve efforts to achieve the city government's climate objectives.

Use

How performance information is used within city government management systems will determine if performance is improved. This statement may seem obvious, yet it is the use of information that seems to perplex many managers. The analytical framework argues that performance information should be the foundation of organisational learning. As Argyris and Schon (1997) argue, learning in organisations has three dimensions: single-, double-, and triple-loop learning are the ideal forms to be adopted by city governments in their use of performance information relating to climate change policies. In Chap. 4, the book outlined the purpose of single-loop learning as the examination of the efficacy of climate initiatives being undertaken by the city. Double- and triple-loop learning requires city governments to analyse lessons from failure and successes, and translate these into more informed decisions.

Observable evidence from the case studies reveals that single-loop learning is the standard practice for each city. As the performance focus of these governments is primarily on outputs, the tendency has been to undertake single-loop learning to identify more efficient ways of achieving climate goals. The achievement of reducing GHG emissions from specified initiatives was the prime focus of the three cities—so far, they have been successful. While taking different strategies, the general approach has been to establish initiatives and set targets with specific time frames to achieve them. Interviews with managers from Copenhagen and Stockholm suggested that a double-loop approach has not been undertaken at this stage. As an example, Copenhagen's targets for emissions reduction has been significantly impacted by national government decisions not to support a congestion charge in the city and to cancel subsidies for electric vehicles. Copenhagen's response has been a single-loop approach that focuses on reducing emissions by raising the existing wind energy target as a component of the energy supply. Double-loop learning would encourage the city government to adopt a wider view and look for initiatives beyond those of the current strategy.

The analytical framework argues that performance information should be used to give an account to internal and external stakeholders of the measurable success of their climate policies. Reporting is a critical use of performance information that reflects a city government's attempts to be transparent and accountable for its climate action. Reports are also an important communication tool to show citizens where success is being achieved and where failure requires new approaches. Transparency and accountability are the key components of building citizen trust in actions of governments responding to climate change. The reports by the case study councils varied in quality and detail. Reports on performance followed the typical communication format of tables, graphs, and charts. There was some narrative, but it was generally brief. Interviewees from each city agreed that current communication formats need to be improved, as they are not appropriate or meaningful. Copenhagen provides both general reports and specific targeted reports designed for different purposes. There has been some experimentation with these reports, as Copenhagen managers argued that they were being designed with specific sectors in mind where action on climate had been slow. Council surveys revealed that 50% of the population of the city were unaware of the climate policies being pursued by the city government. Various forms of reporting and information distribution were under consideration to raise the level of awareness and to promote behaviour change, revealing an attempt to consider more effective use of performance information. Reporting on the broader level however has been inconsistent, with lapses in quality and content. However, the use of information by the Copenhagen Bike Secretariat to understand short trips has been a leading example on the use of performance information to amend, improve, or even replace initiatives to achieve climate objectives (see Box 3.1 in Chap. 6). Centres of policy excellence in various countries are experiments in using evidence to help improve policy. Double-loop learning has proven to be helpful in changing objectives and improving approaches to achieving climate outcomes.

Ideally, reporting against outcomes leads to learning opportunities that raise questions and provide evidence for the changes needed. The reports from the case study cities tend to focus on outputs rather than on outcomes. This is a common weakness in public sector agencies (Marr 2009). Reporting is one of the weakest elements of the measurement activities by the case study city governments. While Copenhagen produced impressive reports on cycling, including stakeholder involvement, its other reports were less comprehensive and inconsistent. Stockholm reports presented results of activities across each of the climate policy areas. These reports provided details of Stockholm's original objectives, a timetable for achievement, annual progress, and comments on progress. Unfortunately, the aggregation of the results into the city government's annual report reduced space for commentary and lost the opportunity to provide a narrative within the organisation and to the community on the challenges and opportunities presented by climate policy action. This is a missed opportunity, as there is clearly, from city council data, strong support for active intervention in environmental issues.

Within each of the reports produced by the three city governments, there are some details provided on the ways in which policy goals are achieved, and there are some details on reasons for lack of achievement. What is missing is any indication of questioning of the original approaches or what next when the targets have been achieved. Both Copenhagen and Stockholm suggest that their strategies are reviewed at regular intervals. Copenhagen's CPH 2025 was reviewed first in 2015–2016. The next reviews take place in 2019–2020 and 2025–2026. The final evaluation will be done through an assessment of the goal-achieving rate (CPH 2025). Stockholm's Climate Action Plan was reviewed in 2016, but at the time of writing (February 2017), it had not been released. These are proposed as major reviews, but details suggest that only single-loop learning is being undertaken and a few new initiatives have emerged from the reviews that re-emphasise and reshape existing initiatives. In contrast, other Scandinavian cities are reassessing the assumptions of their climate strategies by, for example, shifting from incentive-based approaches to examining mandatory approaches, such as a ban on petrol vehicles in their cities to achieve their emissions targets. Research has been quite clear for some time that voluntary approaches targeting production processes appear not to generate significant pollution abatement (Koehler 2007). As mentioned in the book's Introduction, international research finds that there may be a role for mixed approaches and emphasises the complementary potential of voluntary and regulatory approaches to advancing energy efficiency and climate resilience (Trencher et al. 2016). City governments such as Stockholm, Copenhagen, and Tokyo could play an influential role by applying double- and triple-loop learning to many of their climate initiatives. Strengthening of regulatory approaches by city governments needs greater in-depth analysis to reveal potential benefits.

As an example of the impact of a mandatory regulated approach, the Environment Bureau in Tokyo provides some evidence of the contribution of double-loop learning to improve climate outcomes. The original assumptions of Tokyo's climate objectives established in the 2007 *Climate Change Strategy* were reassessed in 2010. In the original plan, building owners were required to provide data on emission levels and energy use. The decision to reduce emissions was left to the building owners on a voluntary basis. The assumption of the city government at the time was that costs would drive the desire for change. When the Bureau's performance information identified a shortfall in achieving the emissions reductions targets, a decision was made to review the overall approach being taken. As a result, the city moved from a voluntary approach to a mandatory cap-and-trade scheme, which has achieved the original target in a shorter time frame.

The *Copenhagen Green Accounts* present an 'overview' of progress on climate and environment-friendly efforts to be carbon neutral in 2025. Details provided in the report cover a broad range of issues the city government believes will interest local residents, such as improvements in cycling conditions, reduction in amounts of CO_2 emissions, improvements to parks and water courses, noise reductions, and organic food availability. While much of the detail does not link directly to the CPH 2025, the city government argues that it provides more comprehensive information on its website, accessible only in Denmark and only in

Danish. Although this is an example of a highly selective use of performance information, to support the image and brand of Copenhagen moving towards carbon neutrality, it reflects a view that the use of performance information is not static, and that different methods can be applied to communicate with local residents that some progress is being made towards the climate goals of the city. As discussed above, Stockholm presents its performance information in an aggregated style within its annual report. The *Stockholm Action Plan for Climate and Energy 2012–2015* surprisingly does not mention any reporting framework. Comments on progress detailed in the annual report are brief and lack explanatory power. As an example, the report on making Stockholm a fossil fuel–free city by 2040 provides minimal details on annual progress, without explaining the overall position in terms of the target (see Fig. 8.1).

This format is typical of many city governments, which represents a top-down approach to reporting that misses an important opportunity for the city to use its performance information to explain to citizens the progress being made, how it contributes to an effective response to climate change, and what more needs to be done. More seriously, such reporting does not reflect a desire by the city to be transparent in its response to climate change. Evidence from the United States and the United Kingdom suggests that the government adopting behavioural insights–based approaches shows the potential role that communication might play in engaging the public in changing their behaviour and possibly reducing their environmental impact (Ockwell et al. 2010; Oullier 2013).

The analytical framework argues that there are benefits in using performance information to benchmark performance both internally against

Indicator	Annual target	Outcome	Goal attainment	Comments
Percentage of land allocations with the maximum energy consumption requirement of 55 kWh/m^2	100%	100%	Met in full	
Green-house gas emissions per Stockholmer (tonne/res.)	3.0	2.8	Met in full	
Electricity consumption per square metre	32 kWh	30.8 kWh	Met in full	
Energy consumption per m^2	160 kWh	158.3 kWh	Met in full	
Proportion of the City's operations that sorts out food wast for biological treatment	55%	65.4%	Met in full	
Number of environmentally classified buildings	35	≥ 60	Met in full	
Amount/proportion food waste that the City's residents and businesses sort out for biological treatment	13.6%	16.5%	Met in full	

Fig. 8.1 Stockholm 2015 annual report on 'climate neutral city'

previous achievements and externally against the achievements of others. Benchmarking provides learning opportunities for policy improvement when performance is contrasted with that of others. Learning opportunities can arise when the performance improvements emerge from policy alternatives or alternative ways of doing the same policy. Observable evidence suggests that there are some efforts by each city to benchmark their performance against other cities in selected areas of activity, for example, bicycle use, public transport, and air quality. Copenhagen provides a Web-based resource called *State of Green Transition Index* which uses performance information to compare the performance of 70 cities across the world in a range of areas, including leadership and climate change, environment, and natural capital. The Index also measures markets and investment, which hints at its real purpose as a marketing tool for high-performing cities such as Copenhagen, Stockholm, Amsterdam, London, and New York, rather than as a tool for policy learning.

According to the analytical framework, performance information should be used to give an account to external stakeholders of the measurable success of their climate policies. Reporting is a critical use of performance information that reflects a city government's attempts to be transparent and accountable for its climate action. Reports are also an important communication tool to show citizens where success is being achieved and where failure requires new approaches. The quality of reports by the case study councils varied in quality and detail. Reports on performance followed the typical communication format of tables, graphs, and charts. There was some narrative, but it was generally brief. Interviewees from each city agreed that current communication formats need to be improved, as they are not appropriate or meaningful. Copenhagen provides both general reports and specific targeted reports designed for different purposes. There has been some experimentation with these reports, as Copenhagen managers argued that they were being designed with specific sectors in mind where action on climate had been slow.

A range of factors, including resource allocation, organisational culture, and what is being measured, determine the use of performance information. Surveys reveal that performance information is likely to be used if senior managers view performance management as important. In city governments, senior managers are more likely to use performance information if the mayor/governor believes performance management is important. In the three cases, senior managers argued that the use of performance information fluctuated over time and with changing regimes. Perhaps the most striking example is the fact that the only comprehensive performance report of Tokyo's 10-year *Climate Change Strategy 2007* was in 2010 under the reign of Governor Ishihara. The three governors since Ishihara resigned in 2012 have not been so enthusiastic about performance reporting. So the Environment Bureau managers have been experimenting with alternative means of reporting progress.

Chapter 4 revealed the central issues that city governments need to consider when accommodating the primary functions of using performance information. There are questions about the intended use of the information, which instruments will maximise the effectiveness of performance to achieve the results being sought, the best data sources for measuring the results intended, the stakeholders to which the reports should be directed, and the most appropriate format to communicate trends, key messages, and outcomes.

Conclusion

According to the majority of climate scientists, the questions about what is happening to the Earth have been settled. The critical question for governments now is how do they respond in effective ways to prevent the scientists' worst fears. This book has investigated the role of city governments and the commitments they are making to do something to effectively respond to the challenges of climate change. The argument of the book is that performance management needs to be a key component of a city government's response to show effectiveness and measurably demonstrate its policy success. The evidence supports two traditional perspectives of performance management. First, performance management has considerable promise in contributing to successful and effective policy implementation. Second, without the application of the fundamental principles established in the performance doctrine, there will be little real impact in achieving climate objectives by city governments. The evidence also suggests that some city governments have been making considerable progress in developing performance systems that demonstrate their effectiveness.

The evidence also demonstrates that much more can, and must, be done before the benefits of performance management are realised. In some instances, greater benefits have been outweighed by other factors entrenched in individual contexts. In other instances, municipal managers have utilised some elements of performance management to work around barriers to change. It seems clear that without fundamental changes to the expectations of city governments, along with organisational reforms that support performance as a driver of change, the full benefits of a performance management approach will not be realised. These changes are all the more important for city governments because of the potential they have to impact on both climate mitigation and adaptation, and because cities have been shown they do better at performance management than other levels of government (Sanger 2008).

City governments are important policy actors in the national and international response to climate change. They operate at the urban scale and are responsible for a substantial portion of government responsibilities that can impact on climate change and how it impacts on citizens. More effective responses will require substantially more engagement from the public to the complex and long-term challenges. Trust is a critical part of the relationship between governments and citizens. Credible information on the behaviour changes required is largely based on the trustworthiness of the source. The effective application of performance management and the principles of the performance doctrine at the city level play an important role in establishing the credibility needed to support behaviour change both within the government and by citizens. A critical lesson from this research is that city governments need to identify the political and organisational factors that may be constraining the successful application of performance management. By focusing on the factors identified in this book, cities will make substantial progress in responding to their climate change agendas of legitimising their role in international negotiations, better engagement with citizens in promoting behaviour change, and embedding climate change considerations in their organisational governance arrangements.

References

Ammons, D. (2015, December). Getting Real About Performance Management: Using Performance Information to Improve Services Is Key. *Public Management*. ICMA Publications, PM Magazine. http://icma.org/en/Article/106489/Getting_Real_About_Performance_Management

Argyris, C., & Schön, D. A. (1997). Organizational Learning: A Theory of Action Perspective. *Reis, 77/78*, 345–348.

Aylett, A. (2015). Relational Agency and the Local Governance of Climate Change: International Trends and an American Exemplar. In C. Johnson, N. Toly, & H. Schroeder (Eds.), *The Urban Climate Challenge: Rethinking the Role of Cities in the Global Climate Regime. Cities and Global Governance* (pp. 156–177). New York: Routledge.

Behn, R. D. (2008). PerformanceStat as a Leadership Strategy: It Don't Mean a Thing if it Ain't Got that Follow-Up. *Twelfth Annual Conference of The International Research Society for Public Management,* Brisbane, Australia.

Book, K., Eskilsson, L., & Khan, J. (2010). Governing the Balance Between Sustainability and Competitiveness in Urban Planning: The Case of the Orestad Model. *Environmental Policy and Governance, 20*(6), 382–396.

Koehler, D. A. (2007). The Effectiveness of Voluntary Environmental Programs—A Policy at a Crossroads? *Policy Studies Journal, 35*(4), 689–722.

Koglin, T. (2015). Organisation Does Matter—Planning for Cycling in Stockholm and Copenhagen. *Transport Policy, 39*, 55–62.

Krause, R. M., Feiock, R. C., & Hawkins, C. V. (2014). The Administrative Organization of Sustainability Within Local Government. *Journal of Public Administration Research and Theory, 26*(1), 113–127.

Loerakker, J. (2013). Story Behind Failure: Copenhagen's Business District Orestad, Failed Architecture. https://www.failedarchitecture.com/the-story-behind-the-failure-copenhagens-business-district-orestad/

Majoor, S. (2014). Ørestad: Copenhagen's Radical New Town Project in Transition. *Planning Theory & Practice, 15*(3), 432–438.

Marr, B. (2009). *Strategic Performance Management in Government and Public Sector Organizations*. Butterworth-Heinemann, Elsevier, London.

McCann, E. (2013). Policy Boosterism, Policy Mobilities, and the Extrospective City. *Urban Geography, 34*(1), 5–29.

Moynihan, D. P. (2008). *The Dynamics of Performance Management: Constructing Information and Reform*. Washington, DC: Georgetown University Press.

Ockwell, D. G., Haum, R., Mallett, A., & Watson, J. (2010). Intellectual Property Rights and Low Carbon Technology Transfer: Conflicting Discourses

of Diffusion and Development. *Global Environmental Change, 20*(4), 729–738.
Oullier, O. (2013). Behavioural Insights Are Vital to Policy-Making. *Nature, 501*(7468), 463.
Parker, R., & Bradley, L. (2000). Organisational Culture in the Public Sector: Evidence from Six Organisations. *International Journal of Public Sector Management, 13*(2), 125–141.
Rainey, H. G. (2009). *Understanding and Managing Public Organizations.* Hoboken, NJ: John Wiley & Sons.
Sanger, M. B. (2008). From Measurement to Management: Breaking Through the Barriers to State and Local Performance. *Public Administration Review, 68*(s1), 570–585.
Sterck, M., & Scheers, B. (2006). Trends in Performance Budgeting in Seven OECD Countries. *Public Performance & Management Review, 30*(1), 47–72.
Trencher, G., Broto, V. C., Takagi, T., Sprigings, Z., Nishida, Y., & Yarime, M. (2016). Innovative Policy Practices to Advance Building Energy Efficiency and Retrofitting: Approaches, Impacts and Challenges in ten C40 Cities. *Environmental Science & Policy, 66*, 353–365.
Van Dooren, W., Bouckaert, G., & Halligan, J. (2010). *Performance Management in the Public Sector*. Hoboken: Routledge.

Index

S. Jones, *Cities Responding to Climate Change*,
DOI 10.1007/978-3-319-64810-1

The manufacturer's authorised representative in the EU is Springer Nature Customer Service Centre GmbH, Europaplatz 3, 69115 Heidelberg, Germany. If you have any concerns regarding our products, please contact ProductSafety@springernature.com

Printed and bound by CPI Group (UK) Ltd, Croydon, CR0 4YY
30/09/2026
02225216-0009